Acne and Acne Scarring Treatments

A Comprehensive Approach

By A. David Rahimi, MD, FAAD, FAACS
Forever Young, Inc. in collaboration with
Karin Olsen

Copyright © 2015 by Dr. A. David Rahimi

ISBN 978-1-329-20372-3

All rights reserved. This book or any portion thereof may not be reproduced or used in any manner whatsoever without the express written permission of the publisher except for the use of brief quotations in a book review or scholarly journal.

First Printing: 2015

Co-written and edited by Karin Olsen
Book design and production by Karin Olsen
AMAZON Marketing Partners / 269 S. Beverly Dr. #750, Beverly Hills, CA 90212 / www.amazonmarketingpartners.com

Forever Young, Inc. / 6333 Wilshire Blvd., Suite 409, Los Angeles, CA 90048
Phone: (323) 653 7700 / www.foreveryoung.net

Acne and Acne Scarring Treatments
A Comprehensive Approach

By A. David Rahimi, MD, FAAD, FAACS
Forever Young, Inc.

Co-written and edited by Karin Olsen

Introduction

Meet Dr. A. David Rahimi

Dr. A. David Rahimi is one of the most experienced cosmetic and acne surgeons in Los Angeles having helped thousands of patients attain optimum skin health and appearance through his dermatology and cosmetic surgery practice.

He treats people of all ages with varying skin colors, types and challenges and, over the last 16 years, has performed thousands of laser treatments, chemical peels and lifts while patients are under local anesthesia. **Yes – local anesthesia.**

A popular doctor who is known for providing holistic patient care rather than lining up a series of procedures, Dr. Rahimi is perhaps even more distinguished by his unique "less is more" approach to anesthesia.

His conservative leanings toward local over general anesthesia started when, as a young resident, he witnessed another doctor's operating room crisis. The scare, which made him question the safety of cosmetic surgery and ultimately his career goals, became the catalyst for how he operates today.

He shares his story here:

After my dermatology residency at Mount Sinai Medical Center in New York, I embarked on a year-long fellowship program to learn the different aspects of cosmetic surgery. I worked with brilliant surgeons from a variety of disciplines including facial plastic surgeons, general surgeons, plastic surgeons and ENT specialists.

I trained in a large and busy outpatient surgical clinic with two operating rooms, two recovery rooms, two hyperbaric oxygen chambers, and 10 examination rooms, staffed by 40 nurses, physician assistants and estheticians. It was a dynamic, immersive and enriching experience during which I learned a great deal about best practices in surgery.

Then, toward the end of my fellowship, I witnessed what was going to become a life-changing event. Another surgeon was preparing to conduct a simple fat transfer procedure when the patient experienced an adverse reaction to the general anesthesia. She crashed and stopped breathing. The emergency medical personnel had to rush into the operating room to re-start her heart.

The next 10 days were among the most difficult of my life. Seeing this young patient on life support fighting to live forever changed the way I approach my work. Miraculously, she fully recovered and I truly believe that her religious faith and the capable doctors at the local hospital saved her life.

After this experience, I had to make a choice. Either give up my goal of becoming a great cosmetic surgeon or find a different and better way of performing surgery.

My salvation came when I learned about Dr. Jeffrey Klein's Tumescent Liposuction procedure. This revolutionary approach to liposuction enables the patient to undergo the treatment while under local anesthesia, minimizing blood loss and the risks of general anesthesia.

In my opinion, Dr. Klein's surgical advancement and leadership regarding this development single-handedly changed the entire field of cosmetic surgery and he forever changed my approach to the practice. I now had the direction I needed to pursue my dream career and help my patients realize their dreams, too.

I'm proud to say, 16 years later, I have built a large practice and have had the opportunity to safely and effectively conduct thousands of cosmetic and plastic surgeries using only local anesthesia with minimal amounts of sedation. It truly has been a dream come true.

Before I detail specific procedures and share before and after photos, I would like to emphasize this book is not a criticism of or attack on any group or society of surgeons. I understand that questioning the status quo could attract criticism of my surgical skills and me as a surgeon. I might be questioned because to some it may appear as if I am criticizing my colleagues. This book is in no way meant to do that.

What this book is meant to do is share information and photos for a variety of procedures I perform on patients under local anesthesia. These procedures include liposuction, fat transfers, face and eyelifts, lasers, chemical peels, and sclerotherapy. I've also included my procedure combination concept Quadrafecta™, which I developed five years ago. Quadrafecta refers to combining less invasive procedures to get even more impressive results with less downtime. I will tell you more about this concept later in the book.

As an aside, please note that while I do not perform breast surgery, tummy tucks, or nose surgery, I have seen such procedures also performed on patients under local anesthesia.

My goal is for patients to understand that they have a choice and can have most, if not all, cosmetic procedures under local anesthesia. I hope that more fellow colleagues will be open-minded enough to give this approach a chance. Please read the book and I look forward to discussing how I can help you achieve your health and cosmetic goals.

About Dr. Rahimi

Dr. Rahimi earned his medical degree with distinction from George Washington University School of Medicine, completed his residency in dermatology at Mount Sinai Medical Center of New York and is Fellowship trained by the American Academy of Cosmetic Surgery.

Credentials

Certified by the American Board of Dermatology and a Diplomat of the American Board of Cosmetic surgery, Dr. Rahimi specializes in minimally invasive procedures that look natural and quickly heal. He advanced the cosmetic facelift with the Tuliplift™, a patented procedure that is safer and more effective than a traditional facelift.

He has also worked closely with medical pioneers such as Dr. Theodore Sutnick, MD who created Manual Epidermal Dermabrasion (MED), a gentler alternative to chemical peels and lasers.

Dr. Rahimi has presented his innovative surgical techniques in numerous published articles and industry conferences. He has received several awards in recognition of his skills, including the Walter F. Rosenberg award in dermatology and the William Newman award in pathology.

He also serves as the co-director of Beverly Hills Cosmetic Surgery Fellowship Program, which is approved by the American Academy of Cosmetic Surgery. The program trains surgeons in the art and science of cosmetic and laser surgery, preparing them to become cosmetic surgeons.

Dr. Rahimi lives in Los Angeles with his wife and three children.

Chapter 1
What is acne?

Acne vulgaris, more commonly known as simply "acne," is an infection of the sebaceous gland and/or hair follicle and mostly occurs on the face, neck, back, chest and shoulders. The sebaceous gland naturally generates an oily substance called sebum which lubricates skin and hair. The overproduction of sebum is the primary contributing factor in developing acne.

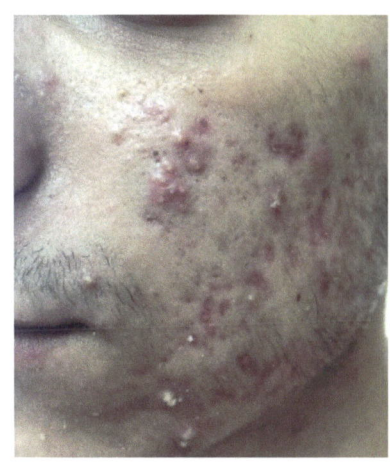

Further aggravating the condition, the sebum in acne sufferers is different as it has higher levels of squalene and wax esters than is typical, as well as lower levels of free fatty acids and linoleic acid. This composition creates an ideal environment for acne-causing bacteria. When the overproduced sebum combines with sloughing dead skin cells, they create a plug that clogs the pore.

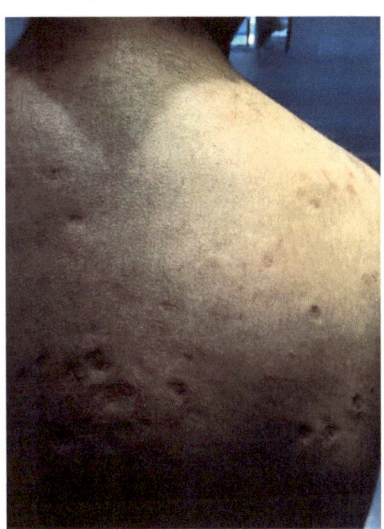

The blocked pore promotes growth of P. Acnes bacteria which thrive in the airless environment causing inflammation and acne ranging from small whiteheads to large, painful and scarring cysts.

While acne is most common in teenagers with a reported prevalence of 70 to 87 percent, it can also be an issue throughout the 20s, 30s and 40s. Adult acne affects 25 percent of adult men and 50 percent of adult women at some time in their adult lives. One third of adults affected with facial acne also have acne on their back and body. For teens and adults, acne can cause depression and social anxiety.

Acne affects all ethnic groups and is more difficult to treat in dark-skinned people. Acne can result from genetics, improper cleansing, and/or regularly touching the area with hands or objects like the phone. Hormonal imbalances and infections can make acne worse. The cost of acne treatment exceeds $1 billion annually in the U.S. alone.

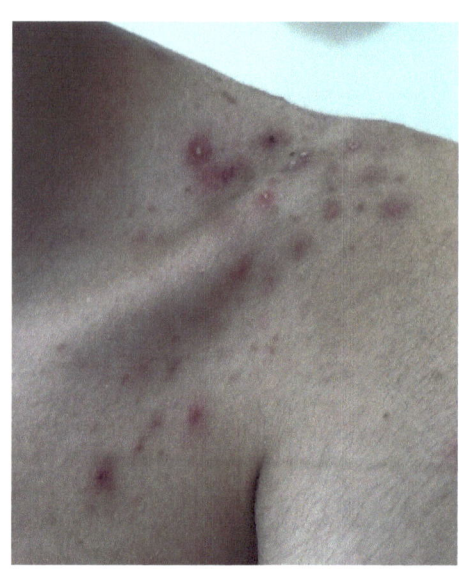

There are six different types of acne:

- **Whiteheads** – Remain under the skin and are usually very small.
- **Blackheads** – Clearly visible, they are black and appear on the surface of the skin.
- **Papules** – Small, usually pink bumps visible on the surface of the skin.
- **Pustules** – Red, pus-filled bumps that are clearly visible on the surface of the skin. Avoid picking as this can cause scars and dark spots on the skin.
- **Nodules** – Large, solid pimples embedded deep in the skin and clearly visible on the surface of the skin. They are often painful. Nodules should be treated by a dermatologist. Over-the-counter treatments may not be strong enough to clear them up, but prescription drugs can be effective.
- **Cysts** – Also known as sebaceous cysts, these are painful, large bumps filled with pus that look like boils and often cause scars. Similar to nodules, cysts can be painful and should be treated by a dermatologist.

Mild acne is generally considered to be fewer than 20 whiteheads or blackheads, fewer than 15 inflamed bumps or fewer than 30 total lesions. It is usually treated with over-the-counter topical medicine and may take up to eight weeks to see a significant improvement.

Moderate acne is generally considered to be 20 to 100 whiteheads or blackheads, 15 to 50 inflamed bumps, or 30 to 125 total lesions.

Dr. Rahimi usually recommends prescription medication for moderate to severe acne. It may take several weeks for patients to notice an improvement and the acne may appear to get worse before it gets better.

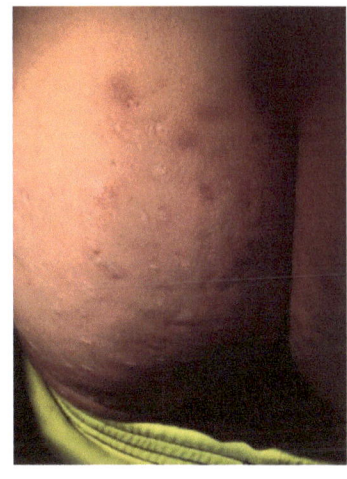

Patients with severe nodulocystic acne have multiple inflamed cysts and nodules. The acne may turn deep red or purple. Dr. Rahimi recommends prompt treatment to minimize the infection so to reduce pain and scarring. One suggested treatment is corticosteroids injections directly into nodules and cysts to reduce the size and inflammation.

Acne conglobata is one of the most severe forms of acne and often leaves scars. It involves inflamed nodules connecting under the skin to other nodules. It can affect the neck, chest, arms, and buttocks. This type of acne is more common in men and is sometimes caused by taking steroids or testosterone. It is important to have a dermatologist treat this type of acne.

Another common variation is acne mechanica which is caused by heat, friction, and pressure against the skin. It is sometimes called "sports-induced acne" because athletes who wear sports gear such as a helmet or baseball cap frequently get this infection. To help prevent it, Dr. Rahimi advises patients wear absorbent materials under sports equipment and cleanse affected areas immediately after activity.

Chapter 2
Acne Causes

Genetics

Genetic makeup is the greatest predictor for determining who gets acne. It has been estimated that people whose parents both had acne have a 75 percent chance of having it and a 50 percent chance if only one parent had it. While genes don't cause acne, they determine the physiology of the skin that increases or decreases the likelihood acne will occur. The over-arching influence of genes in relation to acne ranges from the structure of the hair follicle, to the sebaceous gland activity, to the hormone levels and the immune response to bacteria. Genes also determine sensitivity to dietary and other environmental factors.

Environment

Environmental factors such as weather, topical applications, diet/lifestyle, and medications can also play a big role in developing acne.

Many patients notice the weather tremendously impacts acne as environmental conditions can directly affect the way the body functions. For instance, some find acne symptoms improve during the summer because of higher temperatures and increased exposure to sunlight. On the flip-side, low temperatures may thicken the sebum passing through the follicle and increase the risk of developing a plug.

Cosmetics, skincare and haircare products can also generate and aggravate acne breakouts. To help avoid this, Dr. Rahimi suggests staying away from products that contain lanolin, petrolatum, vegetable oils, butyl stearate, lauryl alcohol and oleic acid.

Diet can also play a major role. Excessive dairy products, meat protein and sugars in the diet have been linked to increased acne and diets low in zinc or high in iodine can worsen acne, especially pustular acne. Smoking has also been linked to acne breakouts and to hindering healing.

Certain medications may also provoke acne. These include:

- Oral corticosteroids, which may also cause steroid acne by increasing yeast proliferation within the hair follicle
- Contraceptive agents such as medroxyprogesterone injection (Depo-Provera), implanted (Jadelle or Implanon) or intrauterine progesterone (e.g. Mirena)
- Oral contraceptives, which reduce circulating sex hormone binding globulin (SHBG) and can sometimes aggravate acne in females
- Testosterone
- Anabolic steroids such as danazol, stanozolol and nandrolone

Hormones

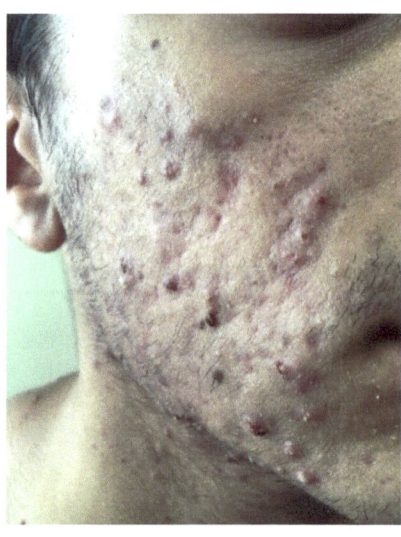

Acne by its very nature can be considered a hormonal disease as hormones are responsible for oil gland maturation and the subsequent capability of these glands to become an acne lesion. To understand the process, start with the previously discussed fact that excessive sebum production is a key characteristic of acne. Hormones enter the picture by stimulating the growth of skin cells called sebocytes, which not only produce the sebum but help usher the sticky substance to the skin's surface. Sebocytes accomplish this by consuming fatty acids to create the sebum and expanding in size until they burst through the hair follicles to spill the sebum onto the skin. Acne results when the sebum becomes blocked in the hair follicle.

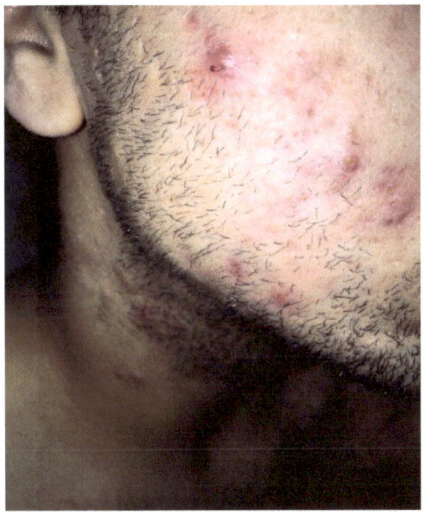

Dead skin cells that meld with the sebum are also a factor in acne. Keratinocytes are the most common cells on the outer layer of skin and they line the follicle walls. In healthy skin, when keratinocytes die they are pushed out by the growing hair. In acne-prone skin, the exfoliation process falters because of excessive skin cell growth and a condition called hyperkeratinization. Keratin is a protein that binds these cells together and too much keratin cements an even stronger bond, making it more likely that dead cells will stick together and block the hair follicle.

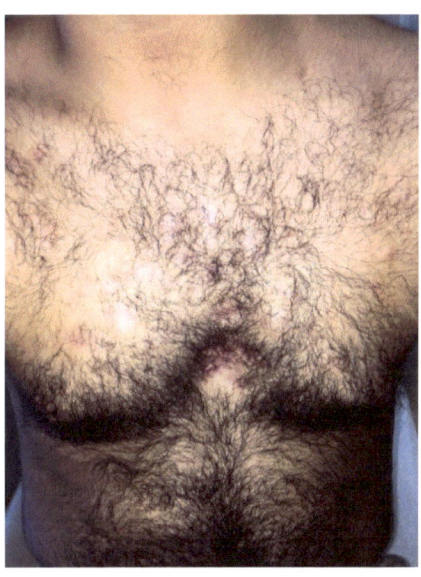

Hormones are at the root of this process as well. Hormones accelerate skin cell growth and it is thought that hormones also increase keratin levels. When this sticky mixture of sebum and dead skin cells blocks the follicle, it starts to swell as more dead skin cells push into the area. The blocked pore causes oxygen levels to sink, creating the ideal environment for P. Acnes bacteria to flourish. When the immune system attacks the bacteria, inflammation ensues resulting in an acne breakout. Acne patients have a distinctly stronger inflammatory response to the bacteria than those with healthy skin and this too is thought to be regulated by hormones, especially in men.

Chapter 3
Acne Scarring Treatments

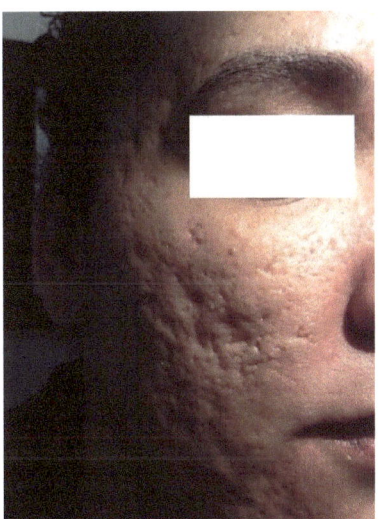

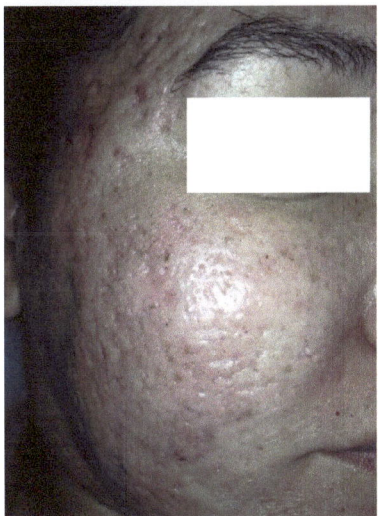

Acne is a complicated disease requiring a customized approach to treatment for each client. Even though the disease primarily stems from hormonal fluctuations that stimulate oil production, not all acne is the same and not all cases of acne will respond to the same treatments. There simply is no one-size-fits-all solution.

Dr. Rahimi has extensive experience treating patients of varying ages with all types of acne and different skin types and colors. Focused on healing the condition, preventing reoccurrence and correcting damage, Dr. Rahimi works with patients to find the best solutions for maximum results for the least downtime and with consideration to individual budgets. In this section, we examine the alternatives in topical and medical medications.

The four major types of medications used to treat acne vulgaris are: keratolytics and exfoliates, antibiotics, retinoids and hormonal treatments. Each class of treatments has pros and cons and Dr. Rahimi weighs these to prescribe the ideal combination.

Keratolytics and Exfoliates

Keratolytics and exfoliates are primarily used to treat mild acne and are not effective in treating moderate to severe cases of the disease. These work to prevent clogged pores by helping to soften, loosen and remove the top layer of the skin. These remedies can also help improve the look of mild acne scars and uneven skin by smoothing the surface and promoting new skin generation.

Keratolytic medications, some of which have antibacterial properties, are available as both prescription and over-the-counter (OTC) medications.

Exfoliates are available in varying strengths ranging from mild products that can be used at home to stronger chemical exfoliates such as chemical peels that must be performed in a professional medical environment.

The OTC chemical peels designed for at-home use are the mildest and least-effective. They are generally low-concentration glycolic acid or alpha hydroxy acid (AHA) based peels that remove a thin layer of the outermost layer of skin. While they may produce some improvement in fine lines and skin tone, they are not strong enough to treat acne scarring.

Stronger peels such as those containing concentrated trichloracetic acid or glycolic acid are for use by professionals only as there is a high risk of permanent skin damage if applied incorrectly.

Dr. Rahimi advises caution regarding excessive use of keratolytic or exfoliate medications as the treatments can cause adverse effects ranging from dry skin to severe and permanent scarring.

A popular alternative to chemical peels is the Apeel MicroExfoliation treatment, which reduces the look of wrinkles and improves skin texture and color by removing the topmost layer of skin. The procedure also stimulates the elastic tissue beneath the surface of the skin resulting in a healthier and firmer appearance. Dr. Rahimi says the NewApeel system is regarded as more gentle and affective than other exfoliating systems, even at higher settings.

Antibiotics

Topical and oral antibiotics help improve acne by killing bacteria. Topical antibiotics, which are applied directly to the affected area, are one of the most common treatments and are used for all types of acne. Oral antibiotics, which are often used to treat moderate to severe acne, are ingested and affect the entire body.

While oral antibiotics can be very successful at treating acne, they have a higher risk of side effects than topical antibiotics. One challenge with the antibiotic category of treatments is the growing issue of antibiotic-resistant bacteria. However, while the resistant bacteria reduce the effectiveness of some antibiotics, there is a wide range of antibiotics from which to choose and Dr. Rahimi will determine which is best for each patient.

Retinoids

Topical and oral retinoids are derivatives of vitamin A and perform a number of functions to prevent and treat acne. First, they inhibit the growth of the sebaceous gland and decrease the production of sebum that feeds bacteria in the hair follicle. This helps decrease bacterial growth in the skin, inflammation and the formation of hyper-keratinized plugs. Retinoids also stimulate cell rejuvenation and are effective in treating mild scars and fine lines.

The topically applied retinoids are commonly used in treating mild to moderate acne. They are often combined with complementary treatments such as an antibiotic and/or a depigmentation agent, such as hydroquinone, for the treatment of hyperpigmented marks. Topical retinoids are limited in their ability to treat severe and cystic acne as they may not effectively penetrate deep into the tissue. Another drawback is they can also irritate the skin, causing increased redness, flaking and general discomfort. However, one strong advantage is they can be applied directly to the affected area therefore avoiding some of the side effects of the oral retinoids.

Oral isotretinoins are members of the oral retinoid family. They are ingested and affect the whole body. Common brands include Accutane, Amnesteem, Isotane, Sotret, Oratane and Roaccutane. They are primarily used to treat moderate to severe cases and are usually reserved for patients who have not responded well to other types of treatments. Isotretinoin therapy is effective as it permanently decreases the size and activity of the sebaceous glands and the physiology of the hair follicle itself. While sebum levels usually return to near-normal levels after completion of the treatment, acne relief often persists for an extended period, and is permanent for many patients.

Out of the common brands, Dr. Rahimi uses Accutane as it is very effective, curing patients with resistant acne 90 percent of the time. However, he uses it as a last resort due to the potential for sometimes serious side effects. Common side effects include transient worsening of acne (lasting two to three weeks), dry lips, dry and fragile skin, and an increased susceptibility to sunburn. Uncommon and rare side effects include muscle aches and pains, and headaches.

Isotretinoin is also known to cause birth defects due to in utero exposure because of the molecule's close resemblance to retinoic acid, a natural vitamin A derivative which controls normal embryonic development. ***It is extremely important to note that women taking isotretinoin must not get pregnant during, and for one month after isotretinoin therapy.*** Sexual abstinence or effective contraception is mandatory during this period. Barrier methods by themselves (such as condoms) are not considered adequate due to the unacceptable failure rates of approximately three percent. Women who become pregnant while undergoing isotretinoin therapy are generally counseled to have a termination. Isotretinoin has no effect on male reproduction.

Hormonal

Many cases of inflammatory acne are instigated by fluctuating hormones. An outbreak occurs when androgen hormones (male hormones) stimulate sebaceous glands and increase sebum production. Overactive sebaceous glands and the resulting excess sebum can contribute to the growth of acne-causing bacteria. The bacteria feed on the sebum resulting in clogged pores.

Treatments that reduce androgen hormone activity can be very helpful in healing all types of acne. One of the most common of these treatments for women and teenage girls is the hormonal birth control pill which contains synthetic estrogens and progesterones (female hormones). Only pills that combine the female hormone estrogen with the synthetic version of the male hormone progesterone (progestin) can stabilize hormonal fluctuations in a way that can treat acne.

There are three oral contraceptives specifically approved by the FDA for acne treatment in women – Yaz, Estrostep, and Ortho Tri-Cyclen. Dr. Rahimi advises patients give the contraceptive pill at least three months to work and says benefits can be seen six months into the treatment.

For post-adolescent acne sufferers, the majority being women, conventional treatments have a high failure rate. These patients do, however, respond well to androgen inhibitors, a class of medications that block androgen hormones by preventing the body from recognizing and/or processing these hormones.

One of these antiandrogenic hormonal therapies proving successful is oral spironolactone (brand name Aldactone). Although spironolactone is actually a blood pressure medication, it has been found to be effective at reducing the appearance of acne either on its own or in combination with oral contraceptives. It works because it blocks aldosterone, the hormone that elevates blood pressure and is chemically similar to testosterone. This likeness means the drug can also block testosterone, reducing androgen levels and with it, sebaceous gland activity and acne symptoms.

A common hormonal disorder caused by androgen imbalance and treated with Aldactone along with birth control pills is Polycystic Ovary Syndrome (PCOS). PCOS affects an estimated five to 10 percent of women of reproductive age (approximately 12 to 45 years old) causing complicated and often serious health problems such as painful ovary cysts, infertility and heart disease.

PCOS can also result in embarrassing and frustrating conditions such as acne, unwanted hair growth and voice change. In PCOS, these disorders appear when the ovaries make excessive testosterone, overwhelming the primary female hormone estrogen and stimulating acne and extra hair on the face and body. In addition to prescribing Aldactone and birth control pills, Dr. Rahimi treats PCOS with a series of other medications including topical creams and oral antibiotics.

Another option is corticosteroid injections which are used to temporarily relieve inflammation associated with severe inflammatory acne characterized by large cysts and nodules.

Chapter 4
Types of Acne Scarring

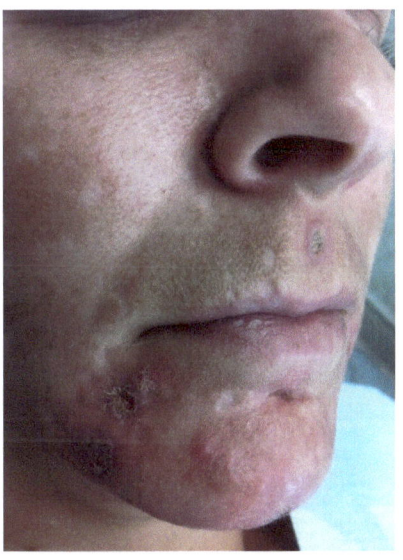

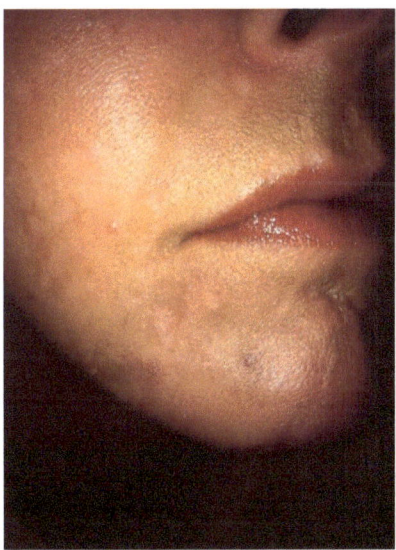

The cycle of acne infection and inflammation can seriously damage the skin and underlying tissue, often resulting in severe and disfiguring scarring. Those who suffer from consistent nodular and cystic acne have the highest risk. During repeated and overlapping outbreaks, the skin has no time to heal and the persistent inflammation and infection prohibit the body from sending the necessary cells and materials for repair. When this happens, the originally healthy tissue is replaced by scar tissue.

While the most effective way to prevent scarring is to treat acne symptoms, there are excellent treatments to diminish the appearance of scarring and drastically improve the skin's overall texture and look. Determining the best treatment, which can range from topical treatments to cosmetic surgery, depends on the type of scarring and the individual patient's skin type and color.

Acne scars appear in different shapes, sizes and colors and each type of acne scar has unique characteristics and features. Generally, acne scars are caused by either a loss of tissue (atrophic) and/or an excess of tissue (hypertrophic). Within these categories, scarring fall into one of three types - depressed (pitted scars), raised (keloid scars) and discoloration scars.

Due to the diversity of the combined factors (underlying causes and types of skin, acne, and scarring) it is crucial to customize treatments, sometimes combining procedures to exact the optimal outcome for each individual patient.

Depressed Scars

The most common type of scar resulting from inflammatory acne is the depressed scar which has three distinct types:

- **Rolling scars** – Caused by long-term inflammatory acne, these are relatively wide and shallow depressions that have rounded, sloping edges. The combination of several of these types of scars clustered together along with the scar-induced remodeled collagen gives the skin a rolling, uneven appearance. Rolling scars tend to become more obvious in aging skin that is losing elasticity and fullness. 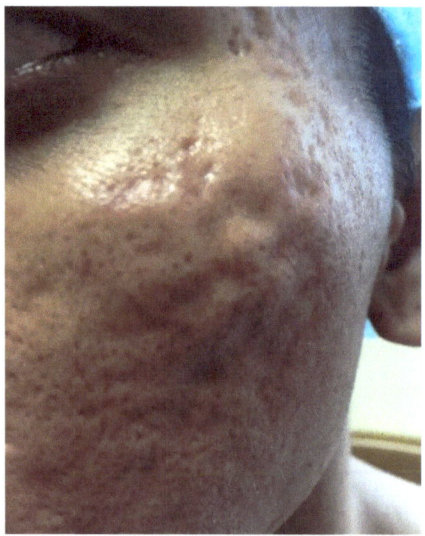 The good news is, because of the sloping borders in the scarred area, many treatments are successful including laser resurfacing, microdermabrasion, Intense Pulsed Light (IPL), needling, chemical peels, and fillers.

- **Boxcar scars** – Often located on the cheeks and temples, these are relatively broad depressions with sharp vertical edges similar to chicken pox scars. They can be shallow or deep. Because box car scars are more angular with steeper edges than rolling scars, it is more challenging to smooth them and blend them into surrounding skin.

To treat the most pronounced scarring, Dr. Rahimi often recommends a series of laser resurfacing treatments and fillers. The shallower boxcar scarring can be treated using chemical peels and/or micro-dermabrasion. Additional options include needling and subcisions.

- **Icepick scars** – These are narrow scars that extend deep into the skin, giving the appearance of having been punctured by an ice pick. In general, depression scars cover a patch of fibrous, collagen-rich scar tissue. This fibrous tissue anchors the scar to the sub-cutaneous tissue, maintaining the depression and preventing the regrowth of healthy tissue. Ice pick scars are often the most difficult to treat without surgery. Because they are so deep, they don't respond to standard resurfacing techniques such as chemical peels, micro-dermabrasion and many types of laser resurfacing. These procedures are unlikely to have significant impact on ice pick scars because they do not remove enough tissue to be effective. Dr. Rahimi's experience indicates some forms of laser therapy such as CO_2, erbium and fractional may be effective at disrupting the underlying scar tissue but he often suggests punch-out excisions as they are mildly invasive and very effective for this type of scarring.

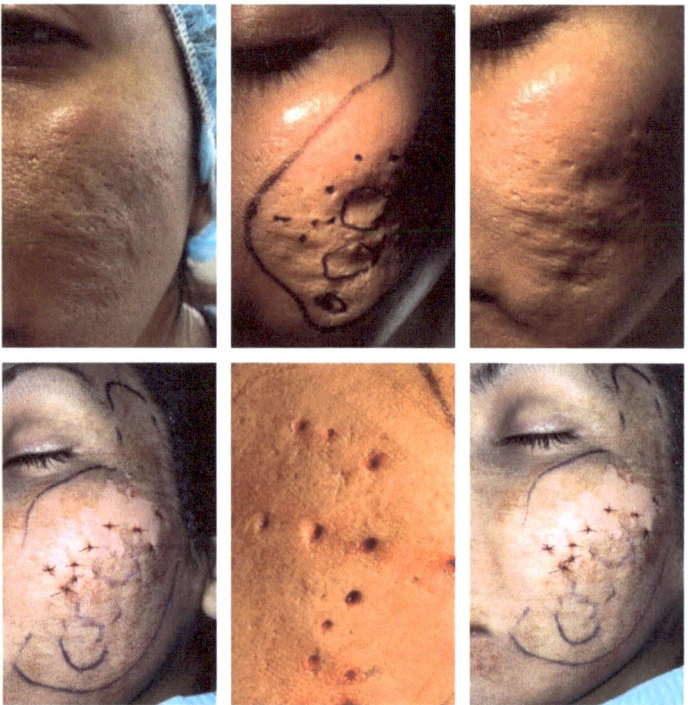

Raised Scars

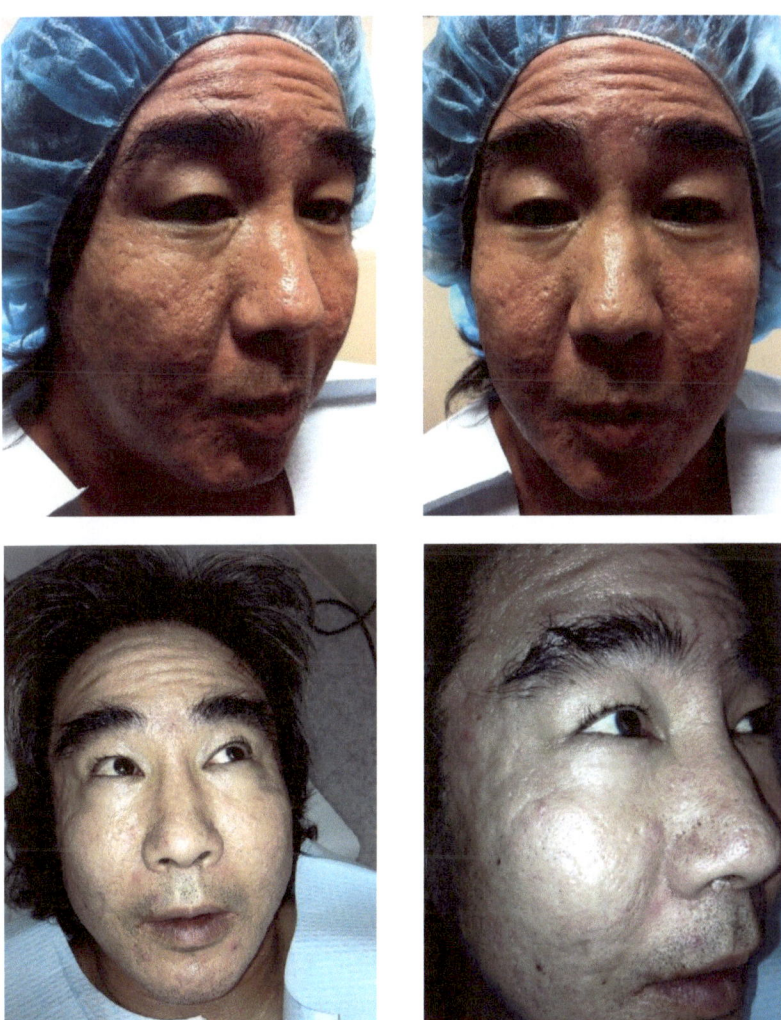

The less commonly occurring raised scar creates a very different set of challenges compared to those of depressed scars. Commonly called hypertrophic or keloid scars, these scars consist of fibrous and firm scar tissue that is often discolored. In its extreme form, the raised scar can grow into a large and dense nodule of rubbery-looking scar tissue.

Due to its elevated nature, this type of scar is best treated by surgically removing or using laser and microdermabrasion procedures. Raised scars generally don't respond to topical treatments such as chemical peels because the bulbous, sinewy tissue is much more resistant to chemical peel ingredients than healthy skin. For those with dark skin, raised scars are more challenging to treat. (Please see paragraph below regarding darker skin.)

Discoloration and Hyperpigmentation Scarring

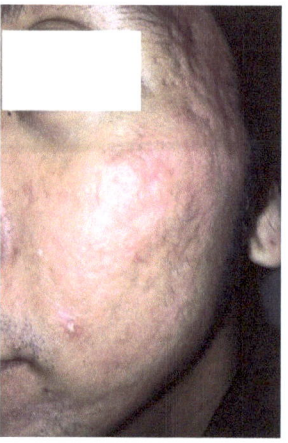

In addition to depressed and raised scarring, traumatic acne damage can also result in long-term or permanent discoloration. Acne scars that cause irregular color in the form of red marks and dark spots are known as postinflammatory hyperpigmentation (PIH).

For hyperpigmentation, increased melanin in the skin darkens an area of the skin in the form of freckle-like spots or blotches. Hyperpigmentation occurs when the cells that produce the melanin pigment called melanocytes begin to gather at the site of an injury or when existing melanocytes begin to produce excess amounts of melanin. Both can happen after any event that causes trauma and inflammation to the skin, such as inflamed acne lesions, abrasions, or burns.

People with darker skin including Hispanic, Asian or African, may be more prone to PIH. The marks range in color from pink to red, purple, brown, or black, depending on skin tone. While they usually will fade over time, the process may take several months and years without treatment.

Fortunately, this type of scarring is the easiest to treat. Dr. Rahimi has attained excellent results using a combination of different treatments including:

- Topical fading lotions (hydroquinone cream, kojic acid, benzoyl peroxide, and licorice)
- Chemical peels including glycolic and salicylic acids, Jessner, and trochloroacetic acid (TCA)
- Laser treatments using Diolite and pulsed dye laser (PDL) to even the color.

Hypopigmentation

Contrasting hyperpigmentation, some scars have hypopigmentation meaning they have less pigment. This is due to melanocytes that are depleted at the injury site or lose their ability to produce melanin. The hypopigmented skin, which has a light, pinkish appearance, often appears in scar tissue and the immediate surrounding area but can also occur in otherwise healthy-looking skin. Hypopigmentation is harder to treat and is most noticeable in people with darker skin tones.

To treat hypopigmentation, Dr. Rahimi uses a light-based device called ReLume that uses flashes of narrow band UVB light to re-pigment light scars. The system's fiber-optic technology allows the light to be both intense and precise. It is a complex procedure requiring five to seven sessions over two years. Areas of lost pigmentation generally respond better when they are new or shallow.

Erythema (Permanent Redness)

Erythema is a condition creating permanent redness in the skin where small capillaries near the surface of the skin are damaged or permanently dilated. It is most visible in light-skin people and is somewhat common in acne patients. Dr. Rahimi's experiences indicate erythema generally responds well to laser and light-based treatments that selectively target hemoglobin. He also attains temporary results treating erythema with topical prescription medications that decrease vasodilation.

Treating Patients with Dark Skin

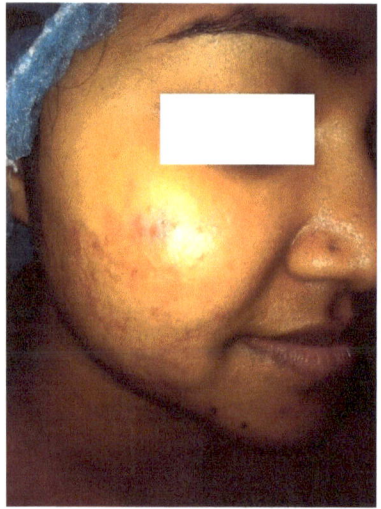
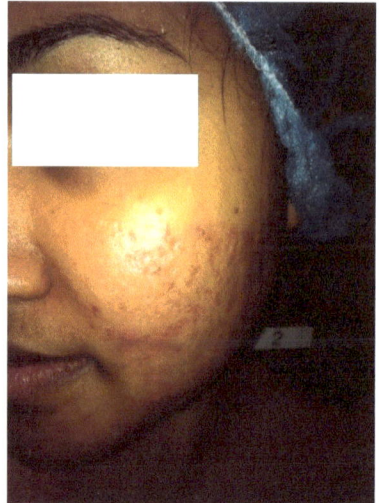

Patients with dark skin who develop inflammatory papules, pustules, nodules and cysts can incur hyperpigmentation, scarring and thick, unsightly scars called keloids.

Acne is common in Asians, who tend to have the PIH dark spots left behind after acne lesions heal. These are challenging to treat.

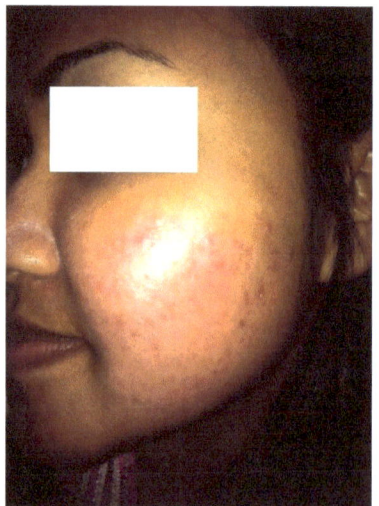

While acne is also the most common skin disorder in African American adolescents and adults, they tend to have inflammatory acne which is easier to treat. Unfortunately, these patients also tend to have post-acne hyperpigmentation and are prone to keloids.

While there has been very little study on acne in the Hispanic population, this segment also experiences a higher occurrence of hyperpigmentation.

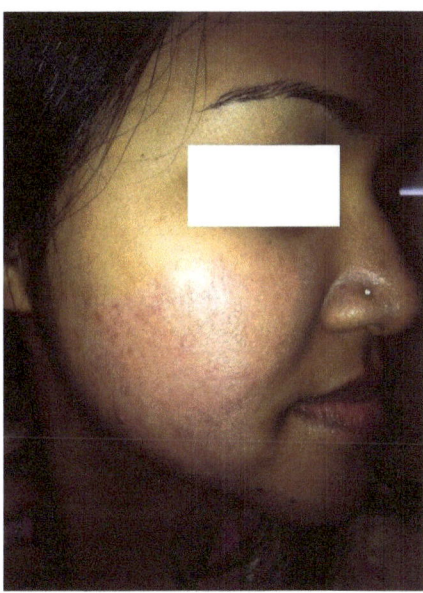

"For skin of all colors, we use a variety of treatment modalities for acne, including TCA peels, Jessner peels, needling, subsicion, punch grafting, stem cell fat grafting, laser resurfacing, dermabrasion and saline injections," Dr. Rahimi says. "Our MiXto laser is an exciting new development in acne scar treatment. This advanced technology delivers powerful results without the harsh side effects and downtime of traditional CO_2 resurfacing."

Chapter 5
Treatments for Acne Scarring

Lasers

CoolTouch CT3Plus Laser

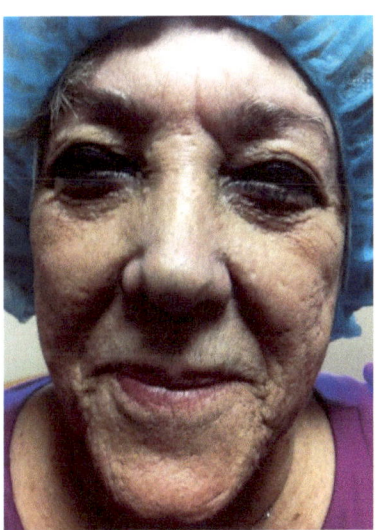

The CoolTouch CT3Plus Laser treatment or CoolBreeze, is a laser treatment that improves active acne and restores skin's natural collagen without scarring or downtime. By targeting those layers of skin where collagen is formed, the CoolTouch laser enables skin to rebuild itself, erasing wrinkles, acne, and acne scarring. This laser is non-ablative, meaning no burning or scabbing.

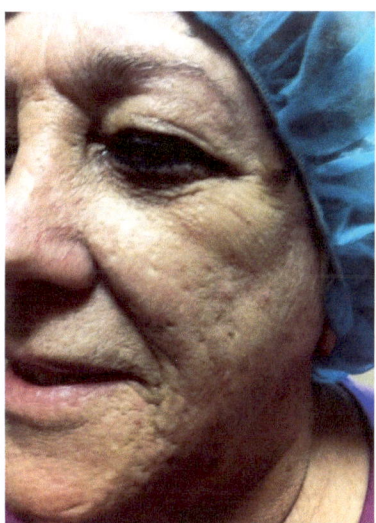

For this procedure, patients are situated in a comfortable and relaxed position and given dark glasses to protect the eyes. Dr. Rahimi uses a special CoolTouch hand piece to deliver the laser light. Before each pulse, he sprays protective coolant on the skin, allowing the laser to penetrate the collagen-producing cells deep in the skin. This targeting ability is unique.

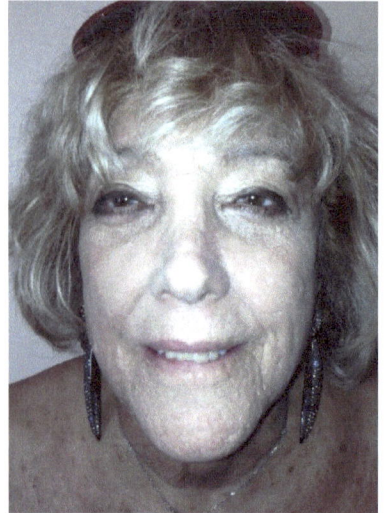 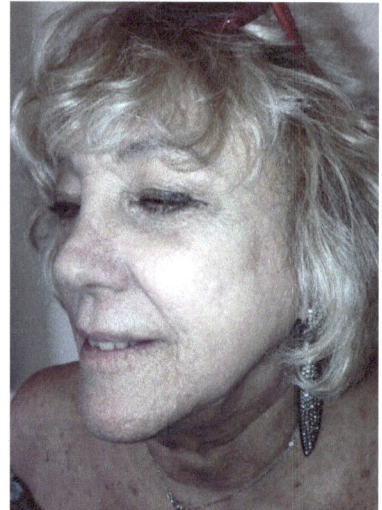

Unlike other treatments that focus on only the bacterial causes, CoolTouch targets acne at its source. The laser penetrates to a level at which it can shrink oil glands, reducing their ability to produce acne, while still allowing for the production of healthy skin oils. Dr. Rahimi fills any acne scarring with the patient's own natural collagen. The treatment takes 30 minutes and there is no downtime.

Patients are given four to six treatments at monthly intervals and will usually see satisfying results after five or six months. To maintain skin improvement, Dr. Rahimi recommends touch-up treatments every month or two.

- **Consultation:** 30 minutes
- **Procedure:** 30 minutes per session / four to six sessions over five to six months
- **Recovery:** no downtime

Iridex Diolite 532 Laser

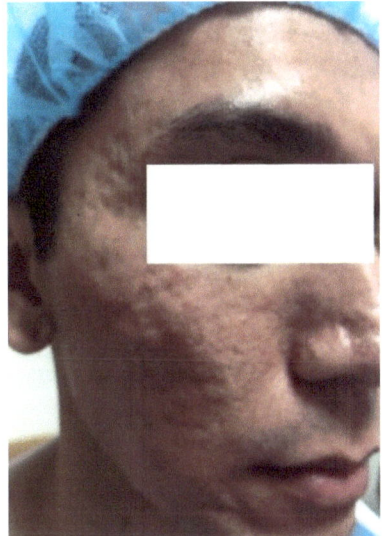

After acne breakouts, the Iridex Diolite 532 laser effectively treats vascular and pigmented lesions including those from blood vessels, brown spots and discolorations from acne scarring.

Following the approximately 15-minute procedure, patients will see improvements almost immediately and there is no post-treatment bruising. While final results should be visible in two to three weeks, in some cases another follow-up treatment may be necessary.

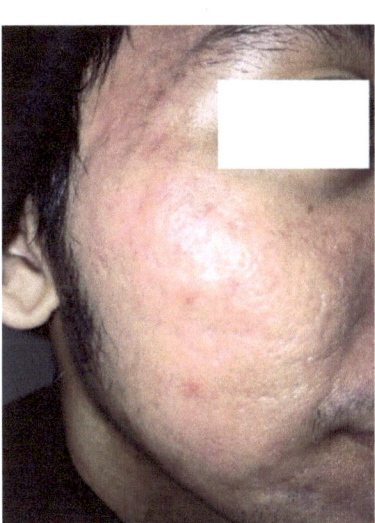

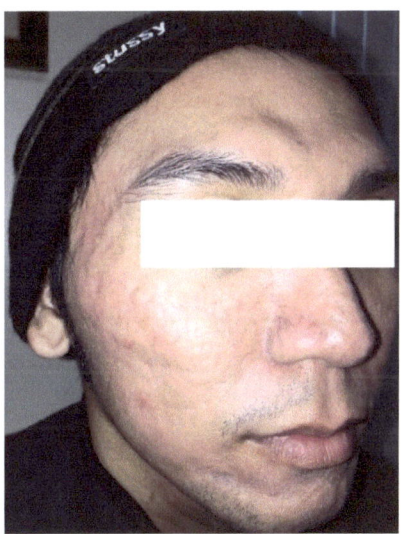

In addition to treating acne-related trouble spots, the laser is also effective for such skin problems as:

- Telangiectasia – spider veins on the face caused by dilation of the capillary vessels
- Rosacea – red lesions around the nose and cheeks, similar in appearance to sunburn
- Spider angiomas – red spider-like veins that are slightly raised in appearance
- Cherry angiomas – red lesions that are slightly raised
- Lentigines – flat brown spots caused by frequent sun exposure
- Dermatosis Papulosa Nigra – small, black marks most commonly found on people of Asian or African descent
- Keratosis – slightly raised spots of pigmentation, found often on the back and hands

- **Consultation:** 30 minutes
- **Procedure:** 15 to 30 minutes
- **Recovery:** one to three days

MiXto SX CO₂ Fractional Laser – The Gold Standard

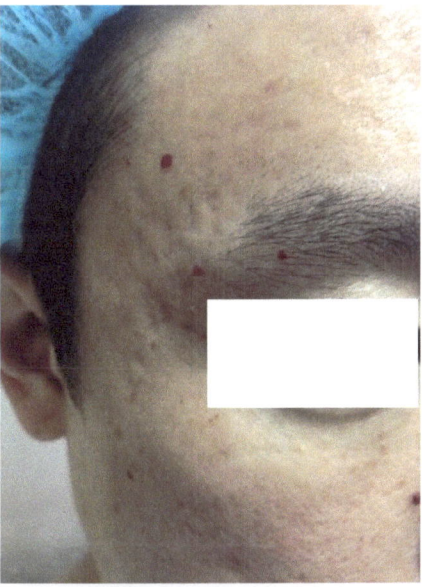

Dr. Rahimi has tremendous success with one of the newer technologies in skin resurfacing, the MiXto SX CO$_2$ Fractional Laser. This laser is ideal for treating some of the most challenging cases of acne scarring and deep pigmentation that have previously been unresponsive to topical agents and peels. The technology is even more impressive considering it provides all the benefits of traditional laser resurfacing in one 30-minute treatment (instead of the traditional three treatments), without general anesthesia and with less discomfort, downtime and cost.

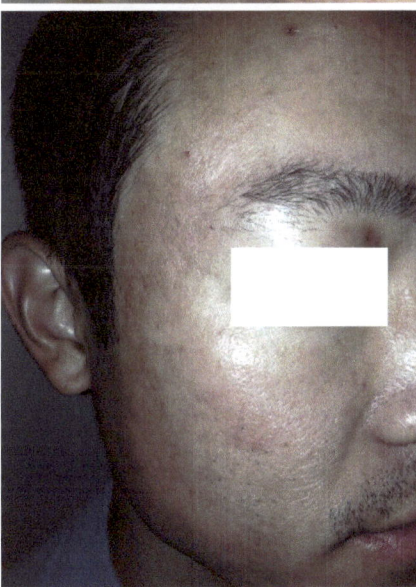

The MiXto SX laser is one of Dr. Rahimi's most sought-after treatments as it is also softens fine lines and wrinkles; removes age, sun spots and pre-cancerous lesions; and spurs collagen production for up to one year following treatment. As an added benefit, the procedure is proven effective for those with darker and ethnic skin.

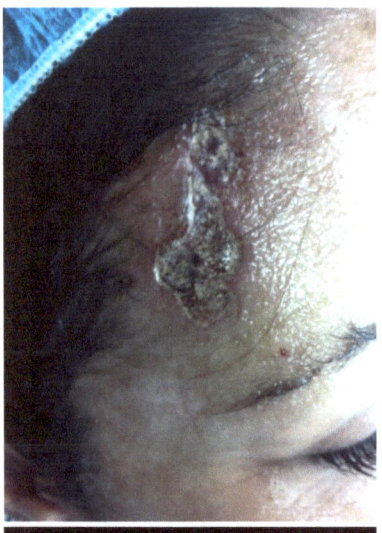

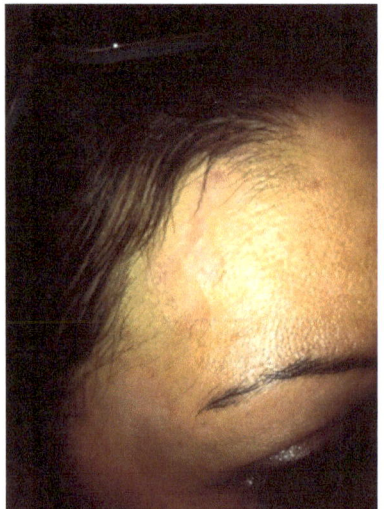

MiXto stands for the mixture between CO_2 energy and fractional resurfacing and the SX stands for surgical because the laser can also be used to make a very clean incision without bleeding.

The MiXto SX laser works by combining the effectiveness of carbon dioxide (CO_2) lasers, long known to be the gold standard in wrinkle removal, with an innovative "quadrant" fractional technology and a more tightly focused beam that delivers uniform tissue ablation to the dermis. This immediately contracts and tightens the damaged skin's tissue as well as improves skin texture and elasticity by producing new collagen over a six to 12-week timeframe.

The CO_2 laser has been used for more than 25 years in the aesthetic industry for treating fine and deep wrinkles, scars, uneven pigmentation and dilated pores. When the CO_2 beam of light comes in contact with the epidermis it heats and vaporizes the skin tissue, instantly removing the superficial layers of the skin, scars and wrinkles while smoothing out the surface of the skin. The skin remodeling occurs with new skin and collagen growth.

Using the MiXto SX laser, Dr. Rahimi has the ability to carefully regulate the depth of penetration and the amount of heat delivered by the beam. The high-speed scanner, driven by a patent-pending algorithm, divides the treatment area into four quadrants or "fractions" and skips the beam from quadrant to quadrant so that each strike is separated by the longest possible interval. The process is repeated until the entire area is treated.

Dividing the laser energy this way allows maximum time for thermal cooling of the tissue, less pain for the patient, and faster recovery. With this "quadrant" fractional technology and a more tightly focused beam, the MiXto SX laser delivers uniform tissue ablation to the dermis causing immediate contraction and tissue tightening.

In spite of the lengthy explanation, the procedure is actually very simple. Thirty minutes prior to the procedure, Dr. Rahimi applies a thick layer of topical anesthesia (EMLA Cream) and administers a small amount of pain medication or Valium to make the procedure more comfortable.

Results become obvious even during the procedure as the skin contracts and shrinks and become more apparent when the skin is completely peeled. Results continue to occur for up to six months as new collagen regenerates and "plumps" up the skin. In most cases, only one treatment is needed, but two or more treatments are beneficial for those with deeper acne scars or wrinkles.

Specifically, the skin is red and swollen the next day from the increased blood flow due to superficial tissue and collagen growth. These symptoms will gradually fade during the first few days. On the second, third and fourth day, there is itchiness similar to a sunburn as the skin begins to shed, revealing a healthy-looking skin. Applying a topical anesthetic will minimize discomfort. Downtime is minimal, usually five days, but most patients are ready to go out in public the following day.

As with all surgical procedures, there are some risks with the MiXto SX but they are minimal. Because only about 20 percent of the skin is "ablated," the risk of scarring and depigmentation are drastically reduced. Additionally, prophylactic antibiotics and anti-viral medication reduce the risk of infection.

In addition to facial skin, the MiXto SX laser can be used on the neck, chest, arms and hands to reduce scarring and discolorations from sun or trauma.

- **Consultation:** 30 minutes
- **Procedure:** 30 minutes
- **Recovery:** One week

Solta Medical Clear + Brilliant Laser

The Solta Medical Clear + Brilliant laser addresses superficial skin imperfections including large pores and brown spots from active acne as well as fine lines. It also helps lift the skin. This laser is ideally suited for patients with darker, ethnic or sensitive skin. Dr. Rahimi has not seen postinflammatory hyperpigmentation or hypopigmentation (light or dark spots) when using this laser.

The Clear + Brilliant laser procedure takes about 20 minutes and is performed after the application of a topical numbing cream. Redness lasts a few hours and patients can return to work immediately. Post-operative care is minimal and includes sunscreen and moisturizers. A series of three to five treatments is needed to achieve long-lasting results.

New skin starts forming within 24 hours but will remain covered by the old superficial layer of the skin. This new skin formation usually lasts for four days but may last longer for some skin types. During that time, normal exfoliation will reveal the repaired skin tissue which is sensitive to the sun and vulnerable to unwanted pigmentation.

To protect the skin, Dr. Rahimi advises patients avoid aggressive exfoliation and frequently apply moisturizers and a dual UVA/UVB sunscreen. It is important that the sunscreen contains both a physical sun block (either or both zinc oxide or titanium dioxide) with a sun protection factor (SPF) of 15 or higher. Avoid direct sunlight and wear sun-protective clothing (e.g. a wide-brimmed hat) when possible.

Most skin care products can be used the day after treatment. Avoid the use of retinoids and topical corticosteroids for one to two weeks before and after treatment. Avoid systemic steroids (e.g., prednisone, dexamethasone) throughout the course of the Clear + Brilliant treatment.

- **Consultation:** 15 minutes
- **Procedure:** 20 to 30 minutes
- **Recovery:** minimal (one day)

Dermabrasion

Surgical Diamond Dermabrasion

Dermabrasion involves removing the top layer of the skin and is most often performed on the face to improve the appearance of acne scars and fine lines around the mouth. While the procedure is ideal for superficial acne scars, deeper scars may require another form of treatment such as punch grafting, elevation, or excision in addition to or instead of dermabrasion.

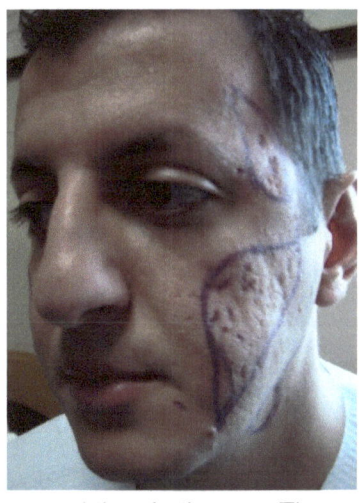

Patients with lighter skin who limit their sun exposure after the procedure tend to have better results than those with darker skin and those who continue to spend unprotected time in the sun. Those who are not good candidates include people who have used isotretinoin in the last six to 12 months; who have had a facelift or brow lift; who have a history of abnormal scarring (keloid or hypertrophic scars); who have an active herpes or other skin infection; who are overly sensitive to cold; or who have a skin, circulation or immune disorder that could affect healing.

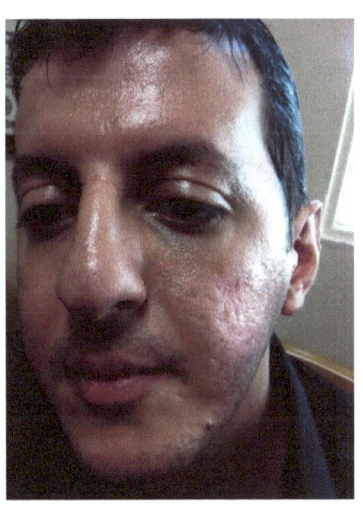

For dermabrasion, Dr. Rahimi uses a rapidly rotating wire brush or diamond wheel with rough edges to remove damaged tissue and level the skin's surface. While the procedure injures the skin and causes it to bleed, the healing process prompts new skin growth replacing the damaged skin. Based on the patient's skin condition, Dr. Rahimi considers brush or wheel coarseness, rotation speed, application pressure and duration to determine the proper treatment depth for optimum results.

To prepare the area, he cleans and marks the skin before applying a local anesthetic to numb the skin. He also applies ice packs to the skin for up to 30 minutes and often uses a freezing spray to harden the skin for deeper abrasions. To manage any possible discomfort, patients can receive additional pain killers and sedation.

Dr. Rahimi treats one small area at a time, applying freezing spray for a few seconds if needed before using the rotating brush or wheel to remove skin. He uses gauze to stop any bleeding before covering with a clean dressing or ointment.

Recovery time depends on the size and depth of the treated area. Generally, skin regrows in five to eight days revealing a pink or red-colored surface which fades in six to 12 weeks. Many patients experience little pain and quickly return to their regular activities. For those who have pain and or swelling, Dr. Rahimi prescribes pain relievers and cortisteroids.

Dr. Rahimi uses surgical diamond dermabrasion as a last resort as the texture of the skin can permanently change. For instance, skin can at times look permanently lighter and/or even glisten and shine. He reserves this treatment for patients with deeper scars that have not seen sufficient improvement with CO_2 laser, PRP, fat grafting or other procedures.

While patients must understand that the scars will improve but the texture may be a little different, Dr. Rahimi says this has not been a tremendous concern for most of his patients. "In the past 16 years I have met many women and a few men who preferred to apply makeup to a smooth, albeit somewhat lighter area, rather than have deeper acne scars."

For more information, please watch the video of Dr. Rahimi performing Dermabrasion here:
https://www.youtube.com/watch?v=BiAyfGPEEJs

Consultation: 30 minutes
Procedure: one to one and a half hours
Downtime and healing: seven to 14 days

Manual Epidermal Dermabrasion (MED)

Manual Epidermal Dermabrasion (MED) is a painless procedure combining physical exfoliation with chemical methods. Safe for any skin type, the 30 to 45-minute procedure promotes collagen production without damaging skin's elasticity and without causing pain or requiring downtime. Dr. Rahimi learned this procedure directly from its pioneering creator, Dr. Theodore Sutnick and has been successfully using the technology for more than a decade.

Dr. Rahimi starts the MED procedure with an expertly detailed superficial skin exfoliation, gently scraping with a scalpel to remove the "dead" layer of skin without leaving a visible mark. He then uses a compressor to extract tiny benign cysts under the skin and superficial rough spots, leaving skin smooth. In the final step, he applies carbon dioxide and acetone to the treated area, causing immediate refinement and tightening. New collagen synthesis begins immediately and continues for weeks.

As a result, the patient's face appears more youthful with significantly diminished fine lines, wrinkles, traces of acne, and pigmentation. Skin color appears fresher and more luminous. To further improve appearance, Dr. Rahimi often combines MED with Botox, Restylane, Perlane, or Hyaline.

In order to achieve consistent and long-lasting results, Dr. Rahimi recommends MED treatments every two to three months. Because MED is extremely beneficial to the skin, Dr. Rahimi recommends patients continue treatments through the years to maintain a flawless complexion.

- **Consultation:** 30 minutes
- **Procedure:** 30 to 45 minutes with additional treatments every two to three months
- **Recovery:** No downtime

Microdermabrasion

Microdermabrasion is a gentle and relaxing treatment that reduces the appearance of age spots, fine lines, and acne scars. Effective for all skin types, the procedure is a non-invasive, non-surgical way to eliminate skin blemishes and leave it feeling soft and fresh. By mixing gentle abrasion with suction to peel off a thin outer layer of skin, microdermabrasion provides immediate results with no downtime.

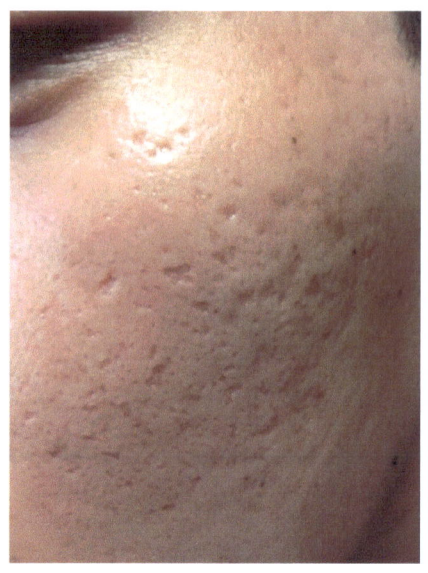

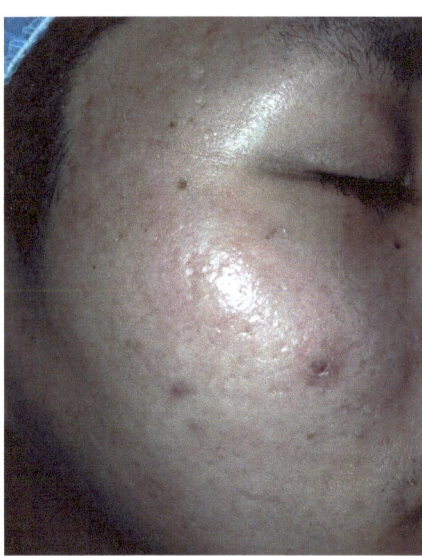

The treatment is a course of 30 to 45-minute weekly sessions during which Dr. Rahimi vacuums micro-crystals through a small hand piece. This microdermabrasion process polishes the skin, removes dead and flaking skin cells, and promotes new skin cell and collagen growth.

Microdermabrasion is a good option for those unable to take time off for healing from chemical or laser treatments or who don't want to interrupt social or recreational activities. It is also excellent for those with sensitive skin that reacts to makeup and chemicals or younger people with early skin changes who want to make early improvements.

- **Consultation:** 30 minutes
- **Procedure:** 30 to 45 minutes, continued every two to four weeks
- **Recovery:** Minimal

Chemical Peels

A good, simple way to address resistant facial acne is with a 45-minute treatment combining a laser treatment with a chemical peel.

For this procedure, Dr. Rahimi starts with the sophisticated non-ablative CoolTouch laser before extracting clogged pores. He then administers a 20 to 50 percent glycolic acid peel and applies a non-comedogenic moisturizer. Downtime, if any, is minimal depending on the occasional redness and flaking.

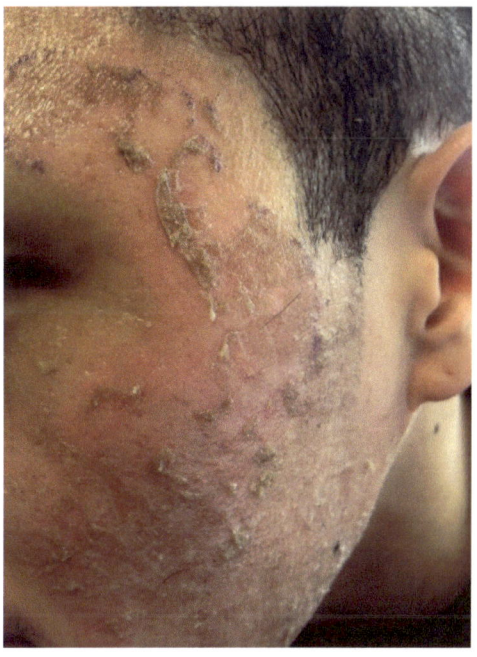

Glycolic or small, regional peels are best for a quick solution. These peels remove the top cell layers which are replaced during the healing process with a fresh, new skin surface. Results take a little longer (five to seven days) but are not as involved as a full chemical peel.

For slightly deeper treatments, Dr. Rahimi administers trichloracetic acid (TCA) and Jessner peels. Both are considered medium-level peels requiring additional sedation. For these procedures, Dr. Rahimi administers a tranquilizer and an oral analgesic. Patients experience only a slight warm or burning sensation.

TCA is a non-toxic chemical, which has been used to perform skin peels for more than 20 years. It is a relative of vinegar (acetic acid). When TCA is applied to the skin, it causes the top layers of cells to dry up and peel off over a period of several days to one week.

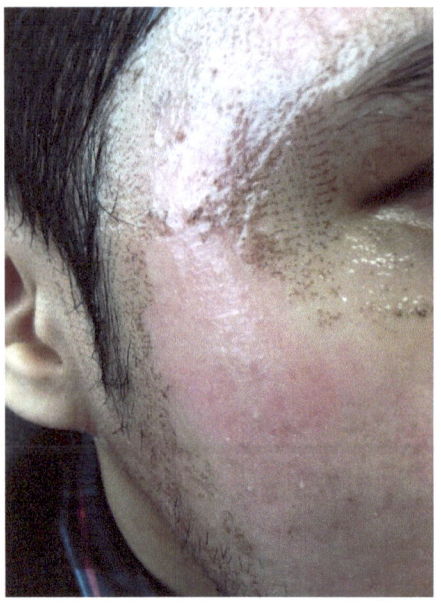

German-American dermatologist Dr. Max Jessner pioneered the Jessner Peel, formerly known as the Coombe's formula. His combination of 14 percent salicylic acid, lactic acid, and resorcinol in an ethanol base is thought to break intracellular bridges between keratinocytes. It is very difficult to "over peel" the skin with the Jessner Peel due to the mild percentages associated with the acid combination and the fact it does not penetrate as deeply as other chemical peels.

As every patient is different and every type of skin reacts differently, Dr. Rahimi generally advises two to three sessions over two months to improve acne and acne scarring.

- **Consultation:** 30 minutes
- **Procedure:** 45 minutes with an additional two to three sessions over two months
- **Recovery:** five to 10 days depending on the strength of the TCA

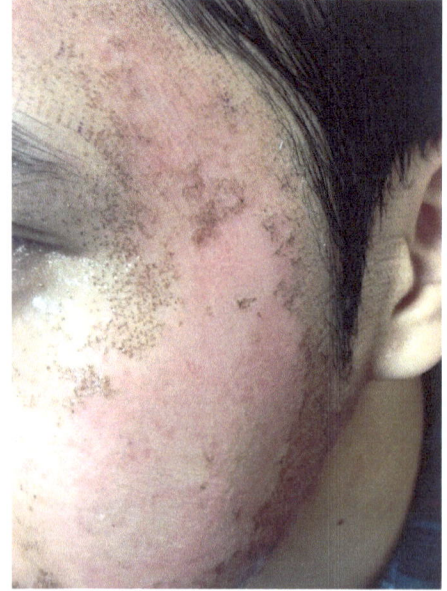

Skin Needling with Dermarollers

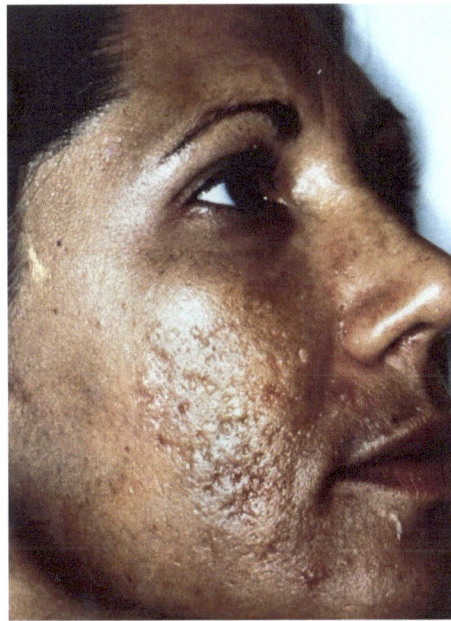

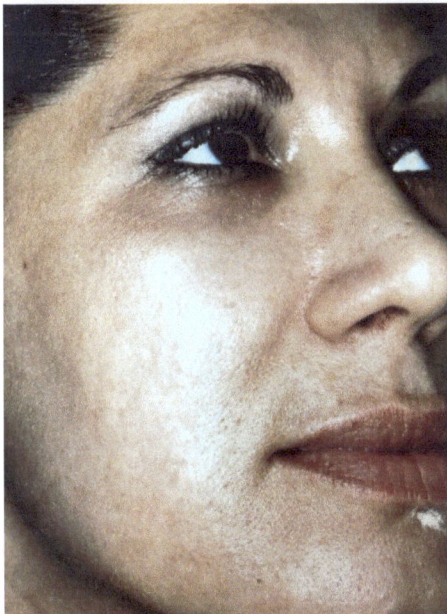

The skin needling or microneedling process induces the formation of deep dermal collagen while reducing the appearance of deep rolling acne scars, wrinkles and uneven pigment. Dr. Rahimi has had excellent results with needling; especially with difficult and depressed acne scars and self-inflicted sores on the face and chest area.

The procedure requires using a special hand-held needling device called a dermaroller, which rather resembles a tiny paint roller. When rolled over the skin, the dermaroller's nearly 200 fine surgical steel micro needles prick the skin, stimulating collagen and helping improve acne scars.

Simply put, Dr. Rahimi says needling is particularly effective because "it fools the body into thinking trauma has occurred at the depths of the rolling acne scars and automatically prompts repair and healing."

Recent trial results published in JAMA Dermatology also support needling as an effective acne scarring treatment.

An excerpt from a Reuters Health article regarding the report stated, "Needling devices that prick the skin with 1mm or 2mm needles on a roller improved acne scars in a recent trial, according to patients and doctors. The patients said the devices improved their acne scars by 41 percent. And dermatologists who were blinded to treatment procedures also reported a statistically significant improvement in scars treated with these devices, researchers said."

Prior to the 30-minute procedure, Dr. Rahimi applies a thick layer of topical anesthesia (EMLA cream) or local injection of lidocaine with epinephrine to significantly reduce any discomfort. He then lightly presses the dermaroller repeatedly over the damaged skin, directing the needles to penetrate the most depressed part of the scar. This action effectively separates the top layer of scarred skin tissue from the deeper scar tissue, drawing small blood droplets under the treated area and stimulating stem cell growth.

While needling may appear to be a simple procedure, Dr. Rahimi stresses there are risks of excessive bleeding, infection and even scarring so it is important a professional performs the needling procedure. "Treating the skin with dermarollers is quite tricky and I have seen both infections and scarring when non-professionals and patients have tried to treat their own skin. It can also be quite painful and result in moderate-to-severe swelling and bruising."

Dr. Rahimi suggests several treatments two weeks apart to get significant improvement and eventually deep rolling scars will level off with the surrounding skin. The needling process can be followed by the application of TCA acids or platelet rich plasma (PRP) and CoolTouch long-wavelength laser treatments to further stimulate the production of collagen and elastic fibers.

- **Consultation:** 30 minutes
- **Procedure:** 30 minutes
- **Recovery:** Minimal with slight redness disappearing in 24 hours

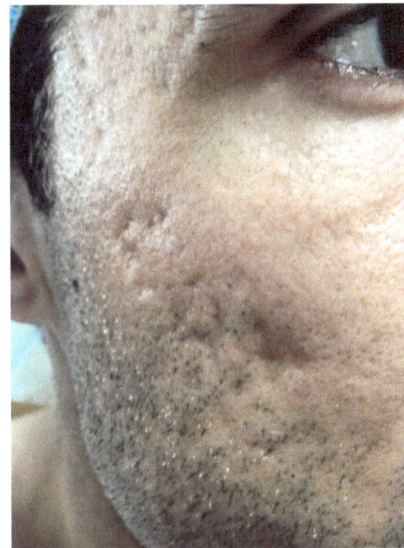

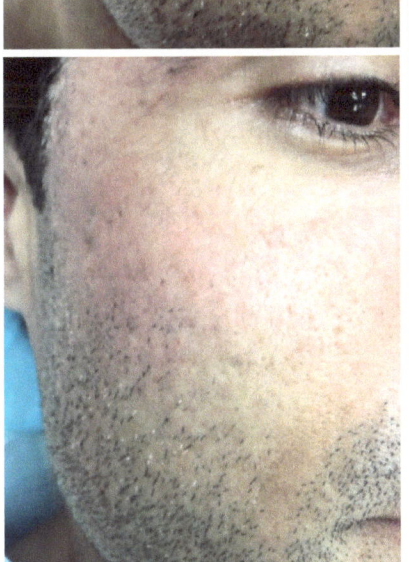

Subcision and Fat Transfers

Subcision and fat transfers are simple and effective procedures Dr. Rahimi uses to improve depressed acne or traumatic scars.

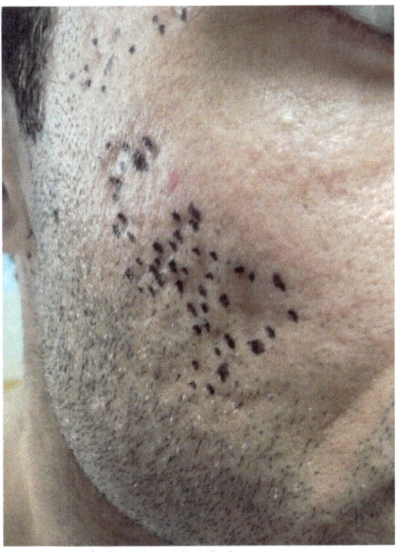

One of the underlying reasons why depressed acne scars are visible is that fibrous bands below the skin's surface pull the skin inward.

To alleviate the pull, Dr. Rahimi numbs the skin with a small amount of local anesthesia (cream or lidocaine solution) and inserts a small NoKor needle under the depressed scar. The NoKor needle features a miniature blade on top of an 18-gauge needle that is so small it doesn't leave a permanent mark.

Then, using a fanning motion in the horizontal direction, he severs the fibrous bands allowing for instant improvement. A few drops of the patient's blood, which contain collagen and elastic fibers, rush into the depression and replace some of the lost collagen.

After severing the fibrous bands, Dr. Rahimi then can further plump up the scarred section of skin by injecting the patient's fat into the area. Injecting the patient's own fat has many advantages including it is a natural filler that is easily harvested and injected into other areas; there are no risks of allergic reaction or rejection by the body; and, once the fat "takes" and the body accepts the fat in the new area, results are permanent.

The procedure is repeated on a monthly basis and can be combined with chemical peels, lasers, or fillers such as Restylane, Juvederm, Perlane, or Radiesse. Downtime is minimal and even patients with sensitive skin and an olive complexion may benefit from the procedure.

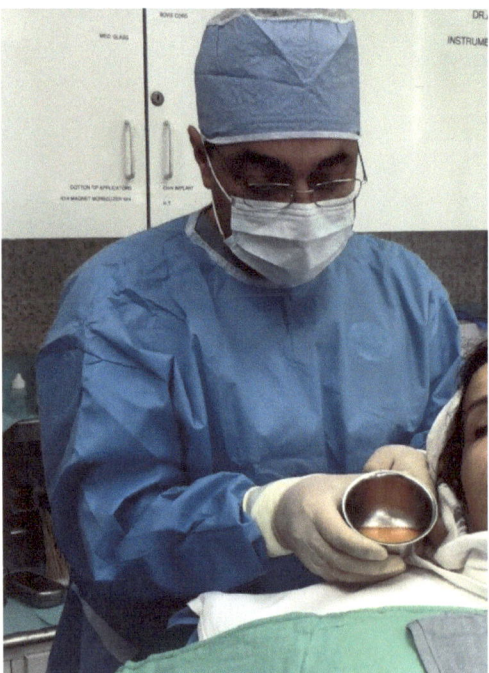

To prepare for the transfer, Dr. Rahimi employs the Coleman technique using special cannulas to harvest the fat from the abdomen on flank area. The harvested fat is spun down and compacted and transferred into 1cc syringes and then injected in micro-droplet format.

Done after laser resurfacing or subcision, the transferred fat prevents the skin from sinking in again and also prevents the deep fibrous bands from re-attaching. The key to success is applying the fat in very small micro-droplets and at different levels. Leftover fat is stored at the clinic's state licensed tissue bank for future use.

The addition of platelet rich plasma (PRP) to the harvested fat has been shown to increase the amount of fat that permanently stays. Please see section on PRP.

- **Consultation:** 30 minutes
- **Procedure:** 30 to 60 minutes
- **Recovery:** Mild swelling and occasional bruising is expected. Infection and irregular absorption and contouring are risks but are minimized by using fine cannulas and careful planning.

Platelet Rich Plasma Treatment for Acne Scarring

Dr. Rahimi has used platelet rich plasma (PRP) injection treatment for acne scarring since 1999. PRP has been used for many years to prompt healing in joints in orthopedic and plastic surgery.

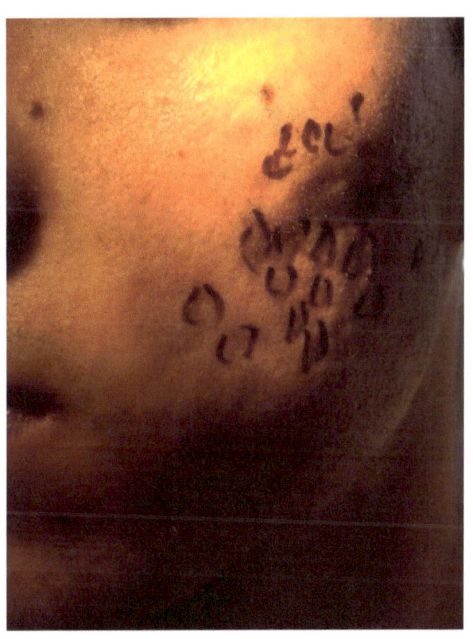

It has also been used as a clinical tool for other treatments including for nerve injury, tendinitis, osteoarthritis, cardiac muscle injury, bone repair and regeneration and oral surgery. PRP has garnered media attention when used to treat professional athletes' sports injuries.

PRP essentially refers to blood that has been distilled down to contain nutrients, plasma, platelets, growth factors and a small concentration of stem cells. The concentrated blood sample contains a high level of platelets (five to 10 times greater than usual) which contain hundreds of proteins called growth factors which is a critical component of the wound-healing process. When injected into scars over several months, PRP drastically reduces scarring, leaving a much smoother skin surface.

For this treatment, Dr. Rahimi uses a Harvest PRP cellular therapy delivery system to develop a concentration of these platelets. After drawing blood from the patient, Dr. Rahimi uses the Harvest system to separate the platelets from the other blood cells. The centrifugation process concentrates the platelets which, when injected into the scarred area, deliver a broad range of growth factors in the proper concentration and ratios to help optimize conditions for healing. Clinical studies have shown that application of PRP can help reduce bleeding, minimize pain, reduce infection rates, and optimize overall healing.

- **Consultation:** 30 minutes
- **Procedure:** 30 to 45 minutes
- **Recovery:** two to three days

Ultherapy, Pellevé, and Thermage

Ultherapy is a new non-surgical, non-invasive procedure that stimulates collagen production. It works by delivering focused ultrasound energy into the skin's deep structural support layers that are typically addressed in cosmetic surgery all without cutting or disrupting the surface of the skin. The ultrasound effect teamed with the body's own natural healing process works to lift, tone and tighten loose skin that has been ravaged by time, sun damage, acne scarring and more.

Ultherapy has become a popular treatment at Forever Young as there's no downtime, no foreign substances, no radical change; just healthy rejuvenation on the inside for a natural, noticeable effect on the outside.

To administer the treatment, Dr. Rahimi maneuvers an Ulthera wand millimeter by millimeter over the treatment area, gliding, pressing firmly, and zapping with the device. After the 30 to 60-minute in-office procedure, patients may notice a short-term "boost" but the natural process of creating new, more elastic collagen builds over time—much like the effect exercise has on building muscle (but without the multiple workouts!)

Results unfold over the course of two to three months and some patients have reported continued improvement for up to six months. While ultrasound does not duplicate the results of surgery, Ultherapy has proven to be an inviting alternative for those who are not yet ready for surgery.

- **Consultation:** 30 minutes
- **Procedure:** 30 to 60 minutes
- **Recovery:** No pain or downtime

Pellevé Nonsurgical Face and Body Tightening

Pellevé is a nonsurgical skin rejuvenation treatment that treats facial wrinkles and other skin imperfections including acne scarring with virtually no pain and no downtime. Likened to a warm facial massage, the treatment can create immediately noticeable results that last.

To administer the 45 to 60-minute Pellevé treatment, Dr. Rahimi uses a hand piece that emits radio frequency energy in the form of electromagnetic waves. This energy heats the dermal tissue below the skin's surface, stimulating collagen production without damaging the epidermis. The new collagen and existing bands of collagen tighten in subcutaneous fat, creating a superficial tightening of the epidermis. As a result, the skin visibly tightens and contours and there is noticeable improvement in skin quality and texture.

There is no need for anesthesia or cooling for this treatment.

- **Consultation:** 30 minutes
- **Procedure:** 45 to 60 minutes per session with a series of three to five sessions for maximum results
- **Recovery:** No pain or downtime

The "Neda" Protocol

Deep acne scarring such as atrophic, ice pick, box car, and rolling is difficult to treat and each type of scar requires a separate modality and special consideration.

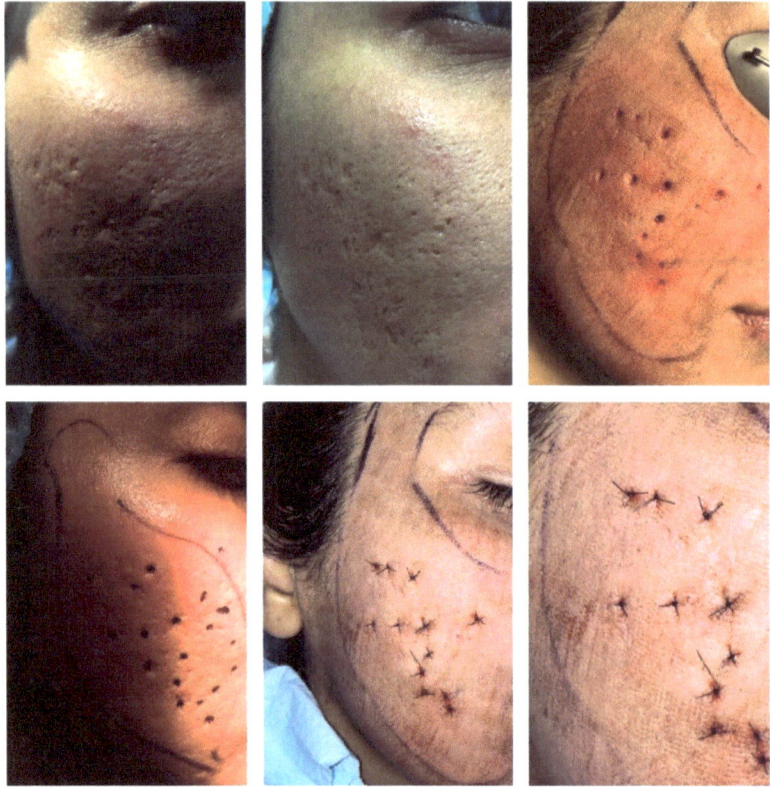

For ice pick scars Dr. Rahimi punches these out using different size round punches. The holes are basically excised and the skin sutured back in place. Dr. Rahimi evens the ensuing excision scar using the MiXto SX laser.

He treats rolling and atrophic scars with subcision using a small knife under the skin to cut the bands that are pulling the skin down. Atrophic areas are filled in with PRP (plasma rich protein) or fat and the skin is tightened using Ultherapy or Pellevé.

Thermage Thermacool and Comfort Pulse Technology (CPT)

Thermage Thermacool is a non-surgical skin tightening treatment that heats the deepest layers of the skin to help tighten existing collagen and stimulate the body's natural renewal of collagen. The treatment's radiofrequency energy tightens the collagen, filling out wrinkles, fine lines, acne scars and other imperfections. The system targets all three layers of facial skin with the goal of keeping the skin rejuvenated and strong for years. Patients generally feel no pain during the treatment and there is little or no downtime.

Dr. Rahimi uses the newest version of the treatment called Thermage Comfort Pulse Technology (CPT) which has proven effective for immediate and long-term results. Thermage CPT is the only FDA approved non-invasive and non-damaging system to use radiofrequency heating combined with external cooling. It is different from previous versions of Thermage as the skin surface is cooled more and each pulse delivers more energy making for a more effective treatment. There is also vibration with each pulse and, between the vibration and the increased cooling, the patient remains comfortable without pain medication.

Dr. Rahimi performs this simple, hour-long procedure with either topical anesthesia or light sedation. He applies radio frequency to the skin and the energy converts into a uniform, sustained deep-tissue heat under the skin. As the collagen rebuilds, the skin and underlying structures tighten and smooth out wrinkles and loose skin. Most patients are able to return to normal activities the following day. A small percentage experience mild swelling and bruising, which resolve within seven to 10 days.

Thermage CPT can also be performed with other procedures such as chemical peels, Restylane treatments and fat transfer.

While almost everyone can benefit from the Thermage system, it is especially suited for patients with a darker complexion, who are not candidates for conventional facelift surgery, and/or those who scar easily.

While the face and neck are the best areas to treat, the arms, legs, breast, thigh, and abdominal areas can also benefit. These areas may, however, require multiple treatments.

While patients notice immediate skin tightening following the Thermage treatment, additional tightening will occur over time with results generally lasting three or more years depending on the patient's natural aging process. Some of Dr. Rahimi's previous patients continue to have excellent, long-lasting results more than three years after the procedure.

According to Dr. Rahimi, the professional administering the treatment is one of the key factors in success. "Sometimes, during the consultation, my patients tell me that they had Thermage several years ago with little visible improvement. Upon further questioning, it becomes clear that the procedure was not performed by a board-certified dermatologist or plastic surgeon. While nurses and nurse practitioners can perform non-surgical lifts, generally speaking, those who aren't medical doctors are more conservative and have less consistent results."

- **Consultation:** 30 minutes
- **Procedure:** One to two hours per session
- **Recovery:** Most patients return to normal activities the following day but a small percentage experience mild swelling and bruising, which resolve within seven to 10 days

Fillers

Fillers instantly smooth away wrinkles, restore natural contours, and improve the texture and appearance of acne-scarred skin. Over the past 10 years the list of facial fillers has grown from one, namely collagen, to more than a dozen FDA approved products.

Dr. Rahimi utilizes the top-performing of these fillers including those especially effective for treating acne scarring -- Juvederm Ultra Plus, Juvederm Voluma, Perlane, Radiesse, Restylane, Restylane Silk, and Sculptra. "The truly amazing thing with fillers is immediate results," says Dr. Rahimi. "When chosen correctly, the body will gradually adsorb the filler over six to 24 months and replace it with its own collagen."

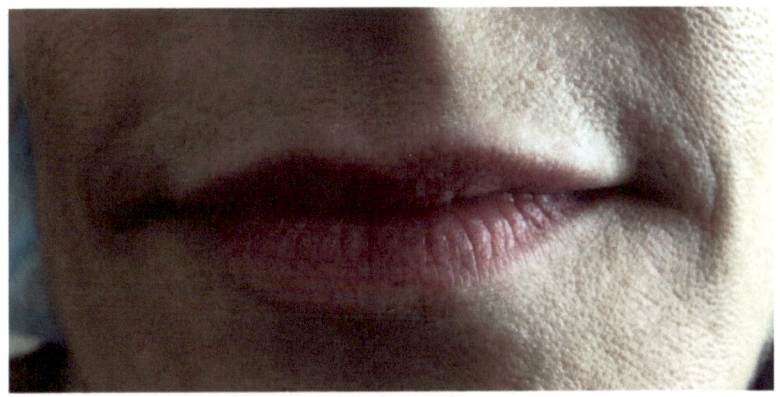

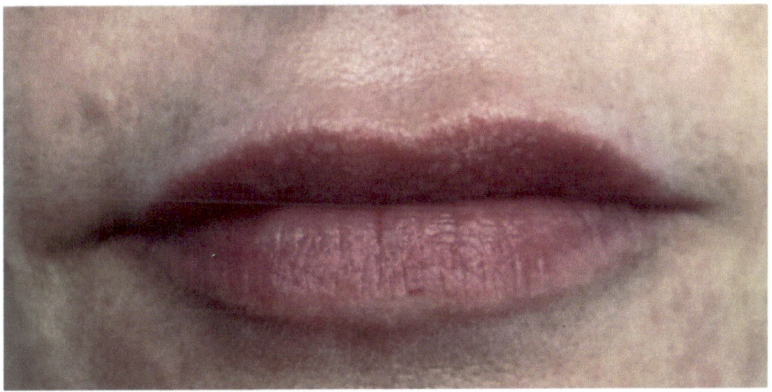

Here's how it works. Young, healthy-looking skin contains an abundance of a naturally hydrating substance called hyaluronic acid (HA). But acne, aging, sunlight and other factors can reduce the amount of HA in skin. The lack of HA causes skin to lose structure and volume, revealing scarring and creating facial wrinkles and folds. Using a dermal filler is a safe and effective way to replace the HA as it fills in the crevices left from acne, builds volume in the scarred areas and smoothes the skin.

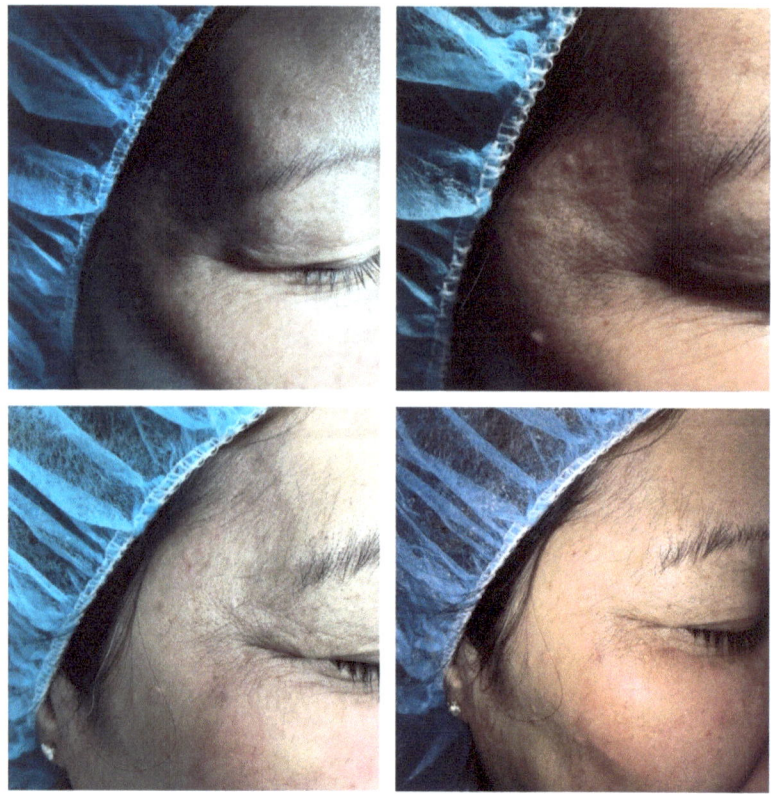

Dermal fillers integrate with the body's collagen and work by replacing lost volume and fullness of the skin and restoring youthful contours. Recent studies show they may also induce neocollagenesis or new collagen by activating certain cells in the skin when injected.

While Dr. Rahimi endorses fillers as a highly effective method to treat acne scarring, he stresses patients should avoid the permanent fillers as "permanent fillers could mean permanent disasters." He explains, "While collagen, hyaluronic acid and fat can be removed if desired, permanent fillers such as silicone and Bio-Alcamid wrap around the nerves and bone and become truly permanent, making it nearly impossible to remove them without cutting that portion of the skin off. Additionally, permanent fillers can also cause bumps and skin thickenings called granulomas that can appear 10 to 15 years later."

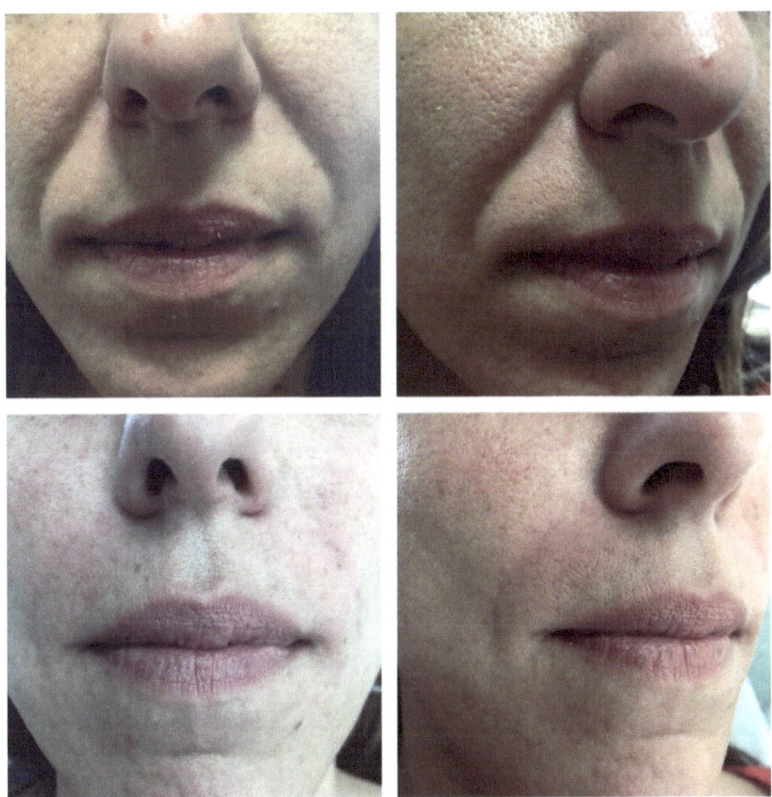

Dr. Rahimi recommends newer fillers such as Juvederm Voluma which generally lasts up to three years and can, over time, have permanent results without the risk of granulomas. Another derma filler with very low risk of forming granulomas is Bellafill, formerly known as ArteFill, which is comprised of 80 percent purified bovine collagen with 20 percent polymethylmethacrylate (PMMA) microspheres, and lidocaine. With the small risk of one in 1000 for developing granulomas, these fillers may be judicially administered in small amounts for select patients in order to correct severe and deep scars.

During the initial consultation, Dr. Rahimi works with the patient to determine which filler is appropriate for the type of skin type and scarring. "For instance, a 'rolling' scar needs different filler than a 'strophic' scar or a very shallow scar," says Dr. Rahimi. "Each filler has certain characteristics regarding thickness and translucency as well as how long it lasts and where it can be inserted. The patient and I discuss the safest and longest-lasting filler for them before proceeding."

Prior to the procedure, Dr. Rahimi suggests patients begin preparing a week before by taking Aspirin and Advil and applying Arnica (available over the counter at most health food stores) to prevent bruising and swelling. For those very sensitive to pain, Dr. Rahimi performs a dental block to reduce discomfort associated with the injections for no extra charge.

For those patients also seeking to reduce the signs of aging, Dr. Rahimi asks patients to bring a photo of themselves from 10 years ago to the initial consultation so he can not only address the scarring, but also refresh the patient's overall appearance.

- JUVÉDERM VOLUMA XC injectable gel is the first and only filler FDA-approved to instantly add volume to the cheek area. It provides a subtle lift, helping to restore contour and a more youthful profile for patients over the age of 21. JUVÉDERM XC smoothes out moderate-to-severe acne scarring, as well as wrinkles and folds around the nose and mouth. It also adds volume to the cheek area, which can further reduce the looks of scarring in this area. JUVÉDERM is manufactured using HYLACROSS technology, creating a smooth-consistency gel. It is infused with lidocaine to improve comfort during treatment.

- The *Restylane* family of products is made using a clear gel formulation of hyaluronic acid — a sugar that is naturally present in skin. *Restylane* works immediately by adding volume to smooth away wrinkles. A clear gel formulation of hyaluronic acid, *Restylane* is specifically formulated to act like the body's hyaluronic acid and eventually naturally breaking down. The product line includes *Restylane*, *Restylane-L*, *Restylane Silk*, *Perlane*, and *Perlane-L*. *Restylane*, *Restylane-L*, *Perlane*, and *Perlane-L* can be used to add volume and fullness to the skin to correct moderate to severe facial wrinkles and folds, such as the lines from the nose to the corners of the mouth (nasolabial folds). *Restylane*, *Restylane-L* and *Restylane Silk* may also be used for lip enhancement in patients over 21 years old.
    - Restylane Silk has recently been approved for release in the United States. It's the first and only FDA-approved product specifically designed for subtle lip enhancement and the smoothing of wrinkles and lines around the mouth in patients over 21 years of age.

- - Like other areas of the face, the lips and skin surrounding the mouth show signs of aging as an individual gets older. This often results in lip thinning, lost shape and an increase in vertical lines above the lip.
  - Restylane Silk is designed specifically to provide natural-looking results in these areas. That's why it's made of smaller, smoother particles than those used in other Restylane products.

- RADIESSE is a derma filler that temporarily adds volume to help smooth moderate to severe facial wrinkles and folds. RADIESSE is made of tiny calcium-based microspheres suspended in a water-based gel.

  The calcium microspheres are similar to minerals found naturally in the body so allergy testing is not required. Once injected, RADIESSE plumps the skin to give the appearance of a smoother surface. Over time, RADIESSE works to stimulate the body to produce collagen naturally. Ultimately the body absorbs the product and leaves behind the natural collagen for long-lasting results. RADIESSE is the only dermal filler available composed of calcium-based microspheres, and has been shown to stimulate the natural production of collagen, for results that may last up to a year or more in many patients.

- Sculptra Aesthetic is a facial injectable that works gradually in a series of treatments — on average, three injection sessions over a few months. It targets the underlying causes of the signs of facial aging and can provide noticeable results that emerge subtly and can last for more than two years. It begins to work within the deep dermis, where the skin's structure is reinforced as Sculptra Aesthetic helps to replace lost collagen. This reinforced collagen structure provides a foundation that gradually restores the look of fullness of shallow-to-deep facial wrinkles. Sculptra Aesthetic is made from a synthetic material called poly-L-lactic acid, which is gradually and naturally absorbed by the body as it works to replace lost collagen. Poly-L-lactic acid has been used for decades in dissolvable stitches and as a facial injectable since 1999 in over 30 countries.

After 24 hours following the procedure, patients should be able to resume normal activities and side effects are moderate and generally last two to four weeks. Common side effects include temporary reactions at the treatment site such as tenderness, swelling, firmness, lumps/bumps, bruising, pain, redness, discoloration, and itching.

- **Consultation:** 15 minutes
- **Procedure:** 15 to 30 minutes
- **Recovery:** 24 hours

Saline Injections

Saline injections are a safe and effective treatment for skin damaged by mild acne and resulting "rolling" or shallow atrophic scars.

To treat damaged skin, Dr. Rahimi injects a sterile water and salt solution strategically into each acne scar. This causes dead cells to release and new cells to form, enabling the skin's connective tissue to return to its natural state.

Although the immediate results are short-term, the repeated applications may permanently change the skin's contour over time. For this reason, Dr. Rahimi recommends patients have saline injection treatments several times over the course of many months.

- **Consultation:** 30 minutes
- **Procedure:** 30 minutes for initial treatment followed by a series of treatments over several months
- **Recovery:** Skin may appear red after the injection but will quickly reduce in intensity to reveal smoother skin

Intense Pulsed Light (IPL) Photorejuvenation

Intense Pulsed Light (IPL) Photorejuvenation is a safe, customizable, non-invasive treatment that uses broad spectrum light to stimulate collagen production. The procedure produces remarkable results in treating a wide range of skin conditions including fine lines, age spots, acne scarring, sun-induced freckles, rosacea and broken capillaries. It can safely treat imperfections on the face, neck chest, arms, and hands. There is no downtime for the procedure as it does not disrupt the skin's surface and it can be combined with other treatments such as Botox and collagen.

To prepare for IPL, Dr. Rahimi applies a cold gel to the treatment area and provides dark glasses to protect the eyes from the bright light. He then glides the IPL hand-piece's smooth glass surface gently over the targeted areas as the pulsating emitted light penetrates deep into the skin.

The treatment is effective because the light converts to heat energy as it reaches beneath the skin's surface, stimulating the body to produce new collagen. During the 20-minute procedure, patients may feel a slight sting similar to rubber band snapping, but generally IPL is painless and rarely requires anesthetic cream.

The treatment is usually performed in a series of four to six sessions. This allows gradual improvement with very low risk and no downtime. Patients are able to return to work the same day.

- **Consultation:** 30 minutes
- **Procedure:** 20 minutes per session per area
- **Recovery:** one to three days for full recovery, but generally there is no downtime as patients are able to return to work the same day

Quadrafecta™

Five years ago Dr. Rahimi developed a combination procedure called Quadrafecta™. For this, he combines four minimally-invasive procedures or devices using local anesthesia to get more impressive results with less downtime. It's an effective and efficient way to address a number of concerns at once and achieve a natural appearance.

During the consultation process, Dr. Rahimi chooses from his more than 20 laser and light sources and, together with the client, creates a customized treatment that can address a number of concerns during one session. These can include tightening and lifting the skin; filling in depressed acne scars and hollowness under the eyes, jawline and more; resurfacing sun-damaged skin; removing excess fat in the jowl and neck area; and replacing lost collagen – all in one visit using local anesthesia.

The following is the list of options considered for the Quadrafecta. While traditionally Dr. Rahimi combines four from the list, occasionally a fifth procedure is needed to address all of the aspects of the aging face.

- Ultherapy (Ulthera) - Ultrasound deep face, eye and neck lift
- Pellevé - Superficial radiofrequency face and neck lift
- MiXto SX CO_2 Fractional Laser - Fractional laser for fine and deep lines
- Fat Transfer with plasma rich protein and stem cells
- Fillers - Sculptra, Radiesse, Juvederm, Restylane, Perlane
- Chemical Peels - TCA and Jessner peels
- Smartlipo Triplex - Fat removal and skin tightening
- IPL Photofacial - Light source to remove brown spots and blood vessels
- Diolite Laser - Remove larger blood vessels
- CoolTouch CT3Plus Laser - Non ablative laser to tighten pores
- Clear + Brilliant Laser - Removes fine line and brown spots
- Thermage CPT - Body lifting with radiofrequency
- Relume - Removal of light scars with focused UVB light
- YAG 5 Laser - Removal of discoloration and irregular pigmentation

American Board of Cosmetic Surgery

American Academy of Cosmetic Surgery

American Society for Dermatologic Surgery

A note from Dr. Rahimi

I have spent more than 16 years compiling the information included in this book. It is meant as a comprehensive guide to the treatment of acne and acne scarring used in my practice. I have omitted a few modalities that I do not perform such as photodynamic therapy, extensive hormonal therapy, and nutritional therapy. I have written about what I know works well to treat acne and acne scarring.

While this guide may seem to offer an overwhelming number of acne and acne scarring treatment options, it is meant to inform and not confuse the patient. As an experienced practitioner, I take many factors into consideration when determining the best treatment approach. These include a patient's skin, past medical and surgical history, and previously used modalities. After analysis, I select from the options detailed in this book to create a customized program for each client.

Please see us for a consultation in our Los Angeles, Beverly Hills, or Westlake Village locations. While many clients reside in Southern California, we also attract patients from around the world including Bora Bora, Spain, Australia, Canada, Japan, Middle East and a dozen states across the country.

If you are interested in a complimentary consultation, please email me a few photos showing the scarring. Include your age, past medical and surgical history, past laser or procedures tried, and specific goals and concerns. I will be happy to send you a plan detailing options that suit your type of scarring. The overview will also include downtime, cost and alternate treatments. Thank you.

afshinrahimi@yahoo.com

A. David Rahimi, MD, F.A.A.D, F.A.A.C.S.

Please download the Forever Young, Inc. app for iPhone and Android devices. This will allow you to watch 40 different laser treatments and surgeries performed by Dr. Rahimi.

American Board of Cosmetic Surgery

AMERICAN ACADEMY OF COSMETIC SURGERY

American Society for Dermatologic Surgery

www.ingramcontent.com/pod-product-compliance
Lightning Source LLC
Chambersburg PA
CBHW041105180526
45172CB00001B/120